Ultramafic Geoecology of North America

Arctic to Caribbean

Earl B Alexander

Soils and Geoecology

2021

Ultramafic Geoecology of North America
Arctic to Caribbean

Alexander, Earl B.
Soils and Geoecology
106 Leland Lane
Pittsburg CA 945-565-5300

ISBN

This book, *Ultramafic Geoecology of North America, Arctic to Caribbean*, is a revision of *Ultramafic Geoecology of North America, Arctic to Tropical* that was printed without color by Lulu Publishing Services in 2020.

1. Earth Sciences. 2. Rocks and Minerals. 3. Landscapes. 4. Ecology. 5.Soils. 6. Plant Communities

Preface

Ultramafic rocks are alien features on land. They are derived from the upper mantle and become components of oceanic plates that drift across the oceans. The plates eventually sink down into the mantle and only minor amounts of them are deposited on continental crust. No more than 1% of the land has exposures of ultramafic rocks.

The chemical compositions of Earth's continental crust and most of the soils on it are much different from that of ultramafic rocks. Plants have evolved to live in soils that have developed on materials that differ greatly from ultramafic materials. Many plants cannot tolerate soils derived from the ultramafic rocks. .Consequently, plant communities on ultramafic soils have unique suites of plants. Some plant species that can cope with the unique chemistry of ultramafic soils grow only in ultramafic habitats where they can avoid competition from plants that do not tolerate ultramafic soils. Because of the unique suites of plants on ultramafic soils, the soils and their habitats are of great interest to ecologists. Understanding the nature and distribution of ultramafic habitats requires some basic knowledge and understanding of the rocks, mineralogy, landscape features, and soils and the distributions of them. The ultramafic landscapes from the Arctic Circle to tropical islands in the Caribbean area afford a broad range of habitats that represent practically all kinds of those on ultramafic soils around the world.

Geoecology is a multidisciplinary science in which the soil and biological features and functions of landscapes are treated as integral systems, geoecosystems. Plants are the dominant biological features. The most productive and abundant plants in geoecosystems have the greatest effects on geoecosystem functions. They are the ones that are given the most attention here.

Many microorganisms that affect plant growth and survival are very important in the functioning of geoecosystems. Until lately, there have been few investigations of microorganisms in ultramafic soils. Brief representations of the bacteria and fungi and their roles are presented where their effects are well known. Lichens can have prominent roles in arctic and subarctic geoecosystems and some that grow at ground level have warranted mention.

Descriptions of the geoecosytems featured in this book are based on the fundamental aspects of ultramafic geology, soils, and plant communities, which are the foundations of ultramafic geoecosystems. Besides introducing basic concepts, the book is an introduction to the diverse ultramafic geoecosystems from the Arctic Circle to hot deserts and to tropical rain forests. Expectations for the *Ultramafic Geoecology of North America, Arctic to Caribbean* book will be fulfilled if it can serve as a foundation for future geoecological investigations of ultramafic landscapes in North America.

References

Alexander, E.B, R.G. Coleman, S. Harrison, and T. Keeler-Wolf. 2007. *Serpentine Geoecology of Western North America*. Oxford University Press.

Acknowledgment

The diversity of subjects in this book is so great that reviewers from a broad range of disciplines were sought for reviews of different chapters. I am pleased to have gotten very helpful reviews from experts in many different disciplines. Among those to whom I am especially grateful are the following:

Chip Bouril, USDA, Natural Resources Conservation Service, California
Robert Boyd, Professor of Botany, Auburn University, AL
Chuck Bulmer, British Columbia Ministry of Forests, Vernon BC
Elena Centeno-García, Instituto de Geología, Universidad Autónomo de México. México D.F.
Stephen Edwards, director emeritus, Regional Parks Botanic Garden, Berkeley CA
Susan Erwin, Natural Resources Specialist, Shasta-Trinity National Forest
Robert C. Graham, University of California, Riverside CA
Lisa Hoover, Six Rivers National Forest, Eureka CA.
Catherine Hulshof, Virginia Commonwealth University
Julie Kierstead, Forest Botanist Shasta-Trinity National Forest
Pavel Kram, Czech Geological Survey
Manuel Matos, USDA, Natural Resources Conservation Service, San Juan PR
Joanne Nelson, British Columbia Geological Survey, Vancouver BC
Richard O'Donnell, retired botanist, California
Nishanta Rajakaruna, College of the Atlantic, Bar Harbor ME
and California Polytechnic State University, San Luis Obispo CA
Jon Rebnam, San Diego Natural History Museum
Samuel Rios, USDA, Natural Resources Conservation Service, Mayagües PR
Paul Sanborn, University of Northern British Columbia, Prince George BC
April Ulery, New Mexico State University, Las Cruces NM
John Wakabayashi, California State University, Fresno CA
Jennifer Wood, Natural Resources Conservation Service, California

Contents

1 Introduction

ultramafic geoecology

the nature and functioning of open systems
comprised of interacting ultramafic ("serpentine")
strata, soils, living organisms, water, and atmosphere

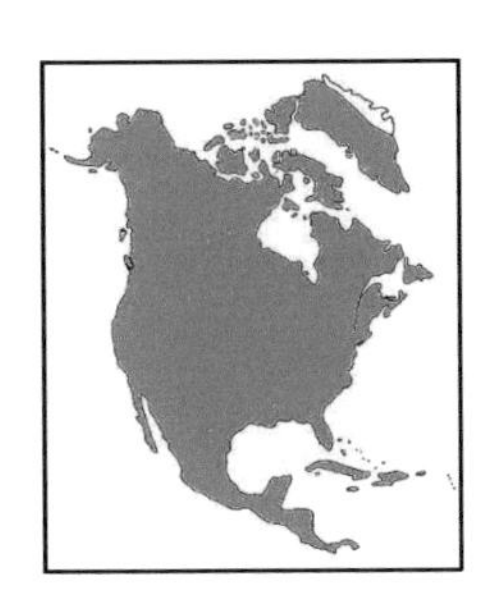

Ultramafic rocks are magnesium-rich, iron-bearing igneous and metamorphic rocks that are sparse components of terrestrial landscapes. They are renowned for their influences on plants and plant distributions. Ultramafic landscapes are of special interest to botanists because of the unique vegetation on them. The vegetative cover is commonly more open in ultramafic areas than in others, and some of the plants may be stunted, never gaining the stature of those in nonultramafic areas. Many kinds of plants grow only in ultramafic areas. Ultramafic areas are fascinating for naturalists, especially for botanists who are keenly interested in the plants that can grow in areas with unique soils..

The unusual effects of "serpentine" on plants are mainly chemical. Ultramafic soils have broad ranges of physical properties, but those properties are not unique among soils. Ultramafic rocks weather to form soils with unique mineral and chemical compositions that depend largely on the mineralogy and chemistry of the parent rocks. Plants get most of their nutrients, other than carbon, from soils. Therefore, a basic knowledge of the geology and soils of ultramafic landscapes is fundamental for understanding the ecology of "serpentine" plant communities.

Landscape systems that include the geologic strata, landforms, soils, and biotic communities are geoecosystems. The *geo* is added to indicate that geoecosystems are complete landscape systems, not more limited ecosystems such as those restricted to the above ground features of plant communities, or to the habitats of particular plant or animal species. Geoecosystems are open systems with air and water entering and leaving them, transporting solids and liquids to and from the systems. The discipline concerning geoecosystems has been called *geoecology* after Troll (1971). It is a discipline that includes the geological, pedological (soils), and biotic aspects of landscapes (Huggett 1995).

An overview of the ultramafic geoecology of North America is presented here, rather than delving into the complexities of a limited number of geoecosystems, in order to establish a broad background for those interested in the ultramafic geoecosystems of North and Central America. Many climatically diverse ultramafic geoecosystems are described from the Arctic Circle to Costa Rica and the Greater Antilles of the Caribbean Sea. These geoecosytems are characterized by the geology, soils, and vegetation in them.

A. Origins and Peculiarities of Ultramafic Rocks

The initial ultramafic rocks are peridotite and dunite that are formed at the top of the mantle, in oceanic settings. Peridotite is commonly hydrothermally altered to produce serpentine and lesser amounts of other minerals that are constituents of serpentinite. Peridotite is ubiquitous beneath the ocean floor but an alien on the continents where vascular plants reside. Peridotite and serpentinite have very high magnesium and relatively low calcium concentrations, making soils derived from them poor substrates for most plants, but some plants thrive on ultramafic soils. Peridotite and serpentinite are commonly called ultramafic rocks, because they have very high magnesium (Mg) concentrations, and large amounts of iron (Fe). Magnesium and Fe are the sources of the letters *maf* in mafic. Water is added in the conversion of peridotite to serpentinite, and some calcium (Ca) can be lost. Peridotite and massive serpentinite can look very similar without magnification to see the different minerals in the rocks. Serpentinite, however, has commonly been stressed tectonically to produce sheared surfaces that make it easily recognizable by smooth surfaces with a blotchy, mottled appearance, generally in shades of green to gray (Fig. 1-1A). The name, serpentine, is derived from *serpentinus* (L.), resembling a snake skin. Ecologists commonly refer to all soils derived from ultramafic rocks as serpentine soils, and plants that are endemic on those soils are referred to as serpentine plants..

Figure 1-1. Ultramafic Rocks. A. Peridotite on the left and serpentinite on the right. The surface of the peridotite is knobby, because olivine has been lost from it by weathering to leave pyroxenes standing out in relief, and iron released from the olivine has been oxidized to produce a reddish brown coating on the peridotite. Serpentinite commonly has a smooth, mottled surface. B. Weathered dunite, composed of olivine and sparse pyroxene has a "buckskin" surface, which is not knobby, because of the paucity of pyroxenes. The hammer handle and its shadow are 27 cm long.

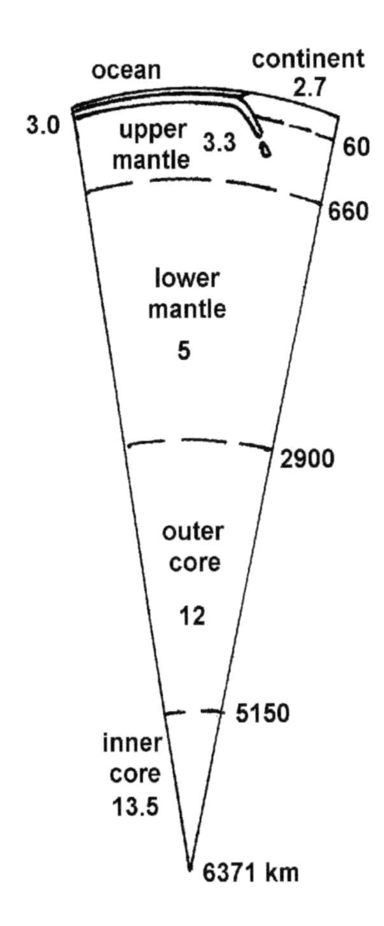

Figure 1-2. A slice of Earth showing the main layers, with the size of the crust exaggerated to show the distinction between oceanic and continental crust. Mean specific gravities are 2.7 in continental and 3.0 in oceanic crust to 13.5 in the inner core.

The chemical elements in peridotite and serpentinite are practically the same, with some addition of water in the hydration of peridotite to produce serpentinite. A big difference between the rocks is in the distributions of the chemical elements among different minerals in them, and differences in the weathering of those minerals, producing peridotite and serpentinite soils with different physical and chemical characteristics. Differences in plant distributions between peridotite and serpentinite soils are generally minor, or illusive, and seem to be related more to physical differences than to chemical differences between the soils.

B. Earth's Ultramafic Character

Earth is composed of three major layers with distinctively different chemical and mineral compositions: core, mantle, and crust (Fig. 1-2). The core, has a radius of about 3471 km, with an inner core composed of solid iron and nickel alloys and an outer core composed of liquid iron and sulfur. Above the core, is a nearly 2900 km thick mantle, which amounts to about 84% of Earth's volume. The lower mantle is composed of iron and magnesium oxides, or oxides and silicates, and the upper mantle is dominated by magnesium and iron silicates, with the proportion of oxides diminishing upward (Condie 2005). Over the mantle lies oceanic crust, which ranges from about 6 to 10 or 12 km thick, and continental crust, which ranges from 30 to about 50 or 70 km thick.

Most primary ultramafic rocks are formed below the oceans where partial melting at the top of the mantle produces a magma that evolves upon cooling to leave depleted peridotite at the bottom, and above that gabbro and basalt, with diabase dikes above the gabbro where magma has flowed upward to feed basalt flows that spread over the ocean floor (Fig. 1-3). Most of these rocks, plus ocean sediments that accumulate on the sea floor, are eventually subducted and recycled back into the mantle. Minor amounts of ocean crust reach the continents and are added to them. On land, the oceanic crust sequence from peridotite through gabbro to basalt is called *ophiolite*. There are several modes of origin for ophiolite (Metcalf and Shervais 2008). Ultramafic rocks are exposed over <1% of the continental surfaces.

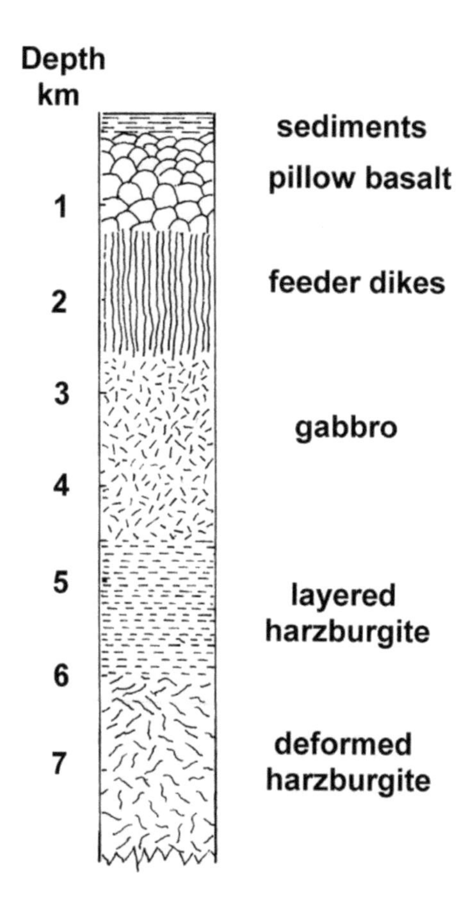

Figure 1-3. The sequence of rocks produced from the magma of partially melted mantle (deformed harzburgite in the figure) at ocean spreading centers. Rocks in this sequence that are transported to continents and added to them are called ophiolites, after they have become continental features.

The elemental compositions of serpentinite and many other ultramafic rocks are similar to that of depleted mantle, which is much different from the composition of Earth's continental crust (Table 1.1). There is much less magnesium (Mg) in continental crust and more calcium (Ca), sodium (Na), and potassium (K) than in the upper mantle. Mass ratios of the Ca and Mg in depleted mantle and continental crust are 0.03 and 1.7, or 0.02 and 1.4 as oxides (CaO/MgO). The molar Ca/Mg ratios \are lower than the mass ratios (Table 1.1); because the molecular weight of Mg is lower than that of Ca, there are more moles of Mg than of Ca in a given mass.

Plants that have evolved in soils with nonultramafic continental parent materials having Ca/Mg ratios >1 g/g have problems growing in soils with ultramafic parent materials having much lower Ca/Mg ratios. Also, potassium and phosphorus, are essential plant nutrients that have low concentrations in rocks derived directly from the mantle (Table 1.1). It is little wonder that relatively few plants have become well adapted to ultramafic rocks and soils with Ca/Mg ratios much lower than one.

Table 1.1 Major chemical elements in continental crust and upper mantle.

Earth	Si	Mg	Ca	Fe	Al	Na	K	P	Ca/Mg	
	oxide of element (average mass %)								mass	molar
Continental Crust	66	2.5	3.6	4.7	14.9	3.4	2.8	0.12	1.7	1.04
Depleted Mantle	44	45.2	1.1	8.2	1.2	<0.1	<0.1	0.02	0.03	0.017

Data from Condie (2005, Tables 2.5 and 4.1). Oxygen is the most abundant element in most rocks of Earth's crust, and in rocks with negligible amounts of sulfides and halides. Thus, the sums of the oxides account for approximately 100% of the elemental compositions in the rocks. The Ca/Mg ratios are elemental, rather than oxide ratios.

C. An Ultramafic Soils Perspective

Ultramafic soils are those that develop in ultramafic parent materials, or in materials weathered from the ultramafic rocks. They range from rudimentary soils in cold climates to deep clayey soils in humid subtropical and tropical climates (Fig. 1-4 and Chapter 12). Ultramafic soils have low Ca/Mg ratios and very high concentrations of the potentially toxic elements chromium (Cr), cobalt (Co), and nickel (Ni). It is the chemical properties of ultramafic soils, rather than the physical properties, that make them unique. Even some highly weathered and leached ultramafic soils that have lost enough Mg to have relatively high Ca/Mg ratios, favorable for most plants, may still lack the plant productivity of most nonultramafic soils.

Compared to the geology and serpentine plant ecology, there has been little interest in the investigation of ultramafic soils. Most soil specialists earn their living by working with the management of forest and agricultural soils, and ultramafic soils are poor for tree growth and agriculture. Consequently, few soils specialists have investigated ultramafic soils in detail.

More than one-quarter of woody plant biomass is in the roots, and more than one-half of grass biomass is generally below ground. Water and all nutrient elements other than carbon, enter plants through their roots. Thus, soils are major components of plant environments and geoecosystems.

The fungi and microorganisms in ultramafic soils have been neglected until recently, although they have been studied intensively for more than a century in other soils. Nearly a century ago, foresters began inoculating seedlings in their nurseries with fungi, because the survival of trees planted without root fungi was poor. Fungi and microorganisms are very important in ultramafic ecosystems. We need to know more about them to have a more comprehensive perspective of ultramafic geoecosystems..

D. Serpentine Plants and Plant Communities

The vegetation of ultramafic soils is generally quite distinctive (Fig. 1-4). The trees are commonly shorter and the vegetation is more open than on nonultramafic soils (Fig. 1-5). These features

Figure 1-4. Soils with serpentinized peridotite parent materials and a great diversity of vegetative cover from the Arctic Circle to Central America. A. Alpine tundra on a moderately deep Inceptisol near the southern margin of the Brooks Range. B. A Jeffrey pine–incense cedar/pinemat manzanita forest on a moderately deep Alfisol in the Lassics area, California Coast Ranges. C. Whiteleaf manzanita–McNab cypress–leather oak chaparral on a shallow argillic Mollisol on Walker Ridge, California. D. A white oak–red maple–hemlock forest on a shallow Mollisol in the Nantahala National Forest, southern Appalachian Mountains. E. A copalquin/palo Adan/sunflower plant community on a duric (silica hardpan) argillic Aridisol in Baja California Sur. F. A Caribbean pine savanna on a haplic Mollisol in Guatemala.

Figure 1-5. An open Jeffrey pine/fescue (*Festuca* spp.) forest on ultramafic soils surrounded by dense Douglas-fir-ponderosa pine/black oak forests on nonultramafic soils. The gray pine trees in the foreground are on a south-facing slope.

reflect the relatively low productivity of ultramafic soils. They are features that are generally obvious to everyone, not only to botanists. Botanists commonly find more endemic plant species on ultramafic soils than on other soils, with the varieties of serpentine plants and the numbers endemic species generally increasing from cold to warmer climates.

Serpentine plant communities, those on ultramafic soils, are generally open, lacking closed canopies of trees (Fig. 1-4B and 1-5). Shrub communities in serpentine chaparral, however, commonly have closed canopies (Fig. 1-4C), or shrub canopies with overlapping crowns. Many alien, or allochthonous, species of plants do not grow on ultramafic soils, leaving the ultramafic habitats to be occupied by the native plants that are more well adapted to ultramafic soils with low Ca/Mg ratios.

Many different strategies have evolved to enable plants to grow in soils with high Mg, Cr, Co, and Ni concentrations (Chapter 4). These strategies include the differential uptake of nutrient and toxic elements from soils, the accumulation of potentially toxic elements in the epidermal and subepidermal tissues of leaves, and the development of tolerance to the toxic elements. Jaffré et al. (1976) found a tropical tree that contained 11.2% (25.7% dry weight) Ni in its sap. Plants that contain more than 0.1% (>1000 μg/g) Cr, Co, or Ni are called hyperaccumulators. Many plants in the Caribbean area are Ni-hyperaccumulators, but there are very few elsewhere in North America. High concentrations of Ni in some plants is a defense against insects that might feed on them (Boyd and Martens 1998). Note: μg/g = ppm (parts per million).

2 Ultramafic Rocks and Geology

Ultramafic rocks come in plutonic, volcanic, and metamorphic varieties and these have different origins and distributions on Earth. Those origins and the compositions of the major kinds of ultramafic rocks are the subjects of this chapter.

A. Earth over Four and One-half Billion Years

Accretion of stellar matter produced Earth by about 4.5 Ga (billion years ago). Earth has been differentiated into a solid inner core of Fe-Ni alloy, a liquid outer core with Fe and S, and a mantle grading from mostly Fe and Mg oxides upward to predominantly Mg silicates with plenty of Fe. Initial intensive bombardment by meteorites continued until about 3.8 Ga. This early period of intense bombardment is called the Hadean, after Hades, the original Greek god of the underworld. Crust that might have formed over Earth during the Hadean was either destroyed by cosmic bombardment or was carried by convection currents downward into the mantle (Condie 2005, Shervais 2006). Radioactive decay of U, Th, and K isotopes generated considerable heat that contributed to active convection in the mantle. The heating by isotopic decay continued following the cessation of heating by Hadean bombardment from meteorites.

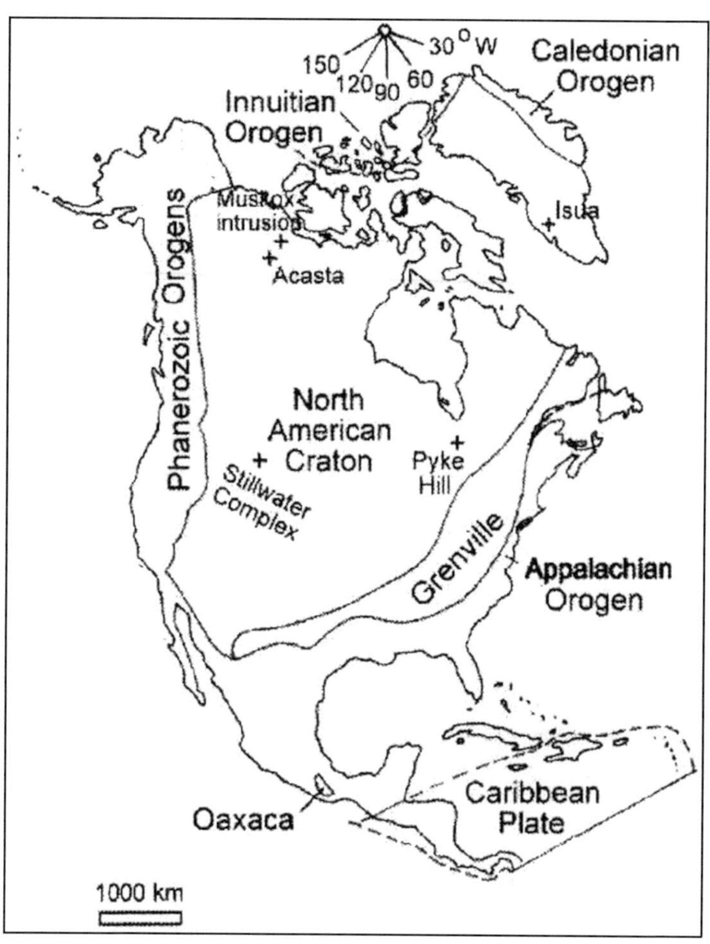

Figure 2-1. Major areas of geological development from the Archean to recent time. Much of the areas of the Paleozoic Appalachian and Ouachitan orogens has been covered by Mesozoic and Cenozoic sediments of the Atlantic and Gulf of Mexico coastal plains.

Zircon crystals 4.2 Ga old have been found in the Acasta gneiss of the Slave Province in Canada (Fig. 2-1), but no rocks of Hadean age have been found, at least no rocks >4.0 Ga. The Acasta gneiss is believed to date from near the beginning of the Archean, about 3.8 Ga. Greenstone belts and TTG (tonalite-trondhjemite-granodiorite) gneisses are characteristic of the early Archean. Some 3.6 to 3.7 Ga greenstones and gneisses, and minor amounts of ultramafic rocks, have been found at Isua in Greenland (Fig. 2-1). Komatiite, along with picrite and picritic (olivine-rich) basalt in volcanic flows of ancient greenstone belts attest to the high temperatures of Archean magma.

Plate tectonics involving the sinking of cold slabs of oceanic crust down into the mantle did not develop significantly until the latter part of the Archean, possibly beginning about 3 Ga (Shervais 2006); and the process was not well developed until the middle of the Proterozoic eon. The main ultramafic rock of the Archean was komatiite. Following the Archean, the dominance of komatiite lava diminished and basalt became more dominant. Some mafic and ultramafic rocks developed as cumulates in sills; sills such as those of the Stillwater complex in the Wyoming Province (Fig. 2-1).

During the middle of the Proterozoic, ancient continental blocks from around the world, including those of the ancestral North American craton, came together and a vast array of exotic terranes were accreted in a sequence of orogenies to create a supercontinent called Rodinia. An area of middle to late Protozoic accreted terranes in eastern North America has been called the Grenville Province. Grenville and like terranes have been found from the Canadian Archipelago and Labrador to Mexico and on all other continents, where they have been given different names. Peridotites and serpentinites from oceanic crust incorporated into Grenville and coeval terranes are found from Canada though the Appalachian Mountains and Texas to Mexico and beyond. The Oaxaca terrane of southern Mexico (Fig. 2-1) is comparable to terranes in the Grenville Province. The amounts of ultramafic rocks are relatively insignificant in the Oaxaca terrane. Archean and Proterozoic provinces of North America from the Arctic Ocean southward into northern Mexico, including those west of the Appalachian Mountains and Grenville terranes that have been covered by Paleozoic and more recent sediments, are referred to as the North American craton (Fig. 2-1).

At about 1.0 Ga, Rodinia. began to break up, and by the end of the Proterozoic or the middle of the Paleozoic crustal plate action began to add terranes to the margins of a precursor of North America (the North American craton) called Laurentia. Laurentia comprised most of the area that is now North America north of latitude 30°N. During the Paleozoic exotic terranes were added around the margins of Laurentia. By the end of the Paleozoic Laurentia and other continents had merged to form a supercontinent called Pangea. The Paleozoic ended at about 0.25 Ga with a catastrophic event, possibly massive volcanism, that caused a major extinction of life that had evolved during the Paleozoic era. During the Mesozoic which followed the Paleozoic era. Pangea began to break up and Laurentia, a component of a larger continent, or supercontinent, called Laurasia, separated from South America and Africa, which were components of another supercontinent called Gondwana.

Following the Paleozoic era, the accretion of peri-cratonic and oceanic terranes have continued on the western margin of North America, but not along the east coast or in the Arctic region. Currently the Yakutat terrane is being accreted along the coast of Alaska.

In the Canadian Archipelago, Neoproterozoic to early Paleozoic pericatonic and oceanic terranes with Laurentian and Baltican affinities were accreted to Laurentia during the late Paleozoic Innuitian (or Ellesmere) orogeny (Fig. 2-1). Minor amounts of ultramafic rocks are exposed in the Pearya terrane, on Ellesmere Island (Estrada et al. 2018)..

A volcanic arc and a large oceanic platform that presumably formed over a large mantle plume developed in the eastern Pacific Ocean during the Mesozoic era (Mann 2007). Subsequently, the arc and platform drifted eastward through a gap that developed between North and South America. Remnants of the volcanic arc are the roots of Cuba, and the Mesozoic ocean platform is now a large part of the Caribbean seafloor.

Of importance to ultramafic geoecology are the kinds of ultramafic rocks and the minerals in them. The geographic distributions of the ultramafic rocks are related to the modes of origin and ages of the rocks. These relationships are outlined in Table 2.1 and discussed in the remainder of this chapter.

Table 2.1 Ultramafic (UM) Rock Types and the Main Minerals in them.

Origin	Age	Location	UM Rock Type	Minerals
top of mantle	Pro, Pal, Mes, Cen	accreted terranes	Dun, Per	Ol, Pyr, Chr
volcanic flow	Archean	Archean cratons	komatiite	Ol, Pyr, Mag
layered body*	Archean	Archean cratons	Per, Pyr	Ol, Pyr, Chr
concentric body	Pal, Mes	Alaska, Brit. Columbia	Dun, Per, Pyr	Ol, Pyr, Chr
hydrothermal alteration	all	all	serpentinite	Ser, Bru, Mag
			steatite, soapst.	Talc, Chl

* Layered bodies in funnel dikes, or lopoliths, and in sills.
Age abbreviations: Pro, Proterozoic; Pal, Paleozoic; Mes, Mesozoic; Cen, Cenozoic
Rock type abbreviations: Dun, dunite; Per, peridotite; Pyr, pyroxenite. Minerals: Bru, brucite; Chl, chlorite; Chr, chromite; Mag, magnetite; Ol, olivine; Pyr, pyroxene; Ser, serpentine.

B. Primary Ultramafic Rocks

Dunite and peridotites, harzburgite and lherzolite (Fig. 2-2), are the major primary ultramafic rocks. Lherzolite is dominant in the upper mantel, with harzburgite above where partial melting has converted much lherzolite to harzburgite, with lesser amounts of dunite.

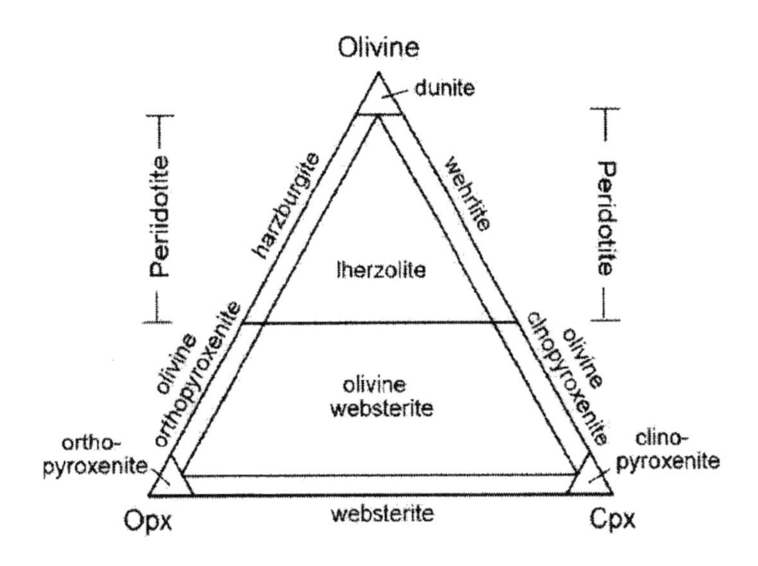

Figure 2-2. Classification of intrusive ultramafic rocks based on their contents of olivine, orthopyroxene (Opx) and clinopyroxenes (Cpx).

The common composition of the upper mantle is that of garnet-lherzolite, or at the more shallow depths, spinel-lherzolite. The main minerals in lherzolite (Fig. 2-2) are olivine ($MgSiO_3$), with some Fe substituting for Mg, and pyroxenes, both orthopyroxene (Mg_2SiO_4), and clinopyroxene ($MgCaSiO_4$), generally with less Fe in pyroxenes than in olivine. Garnet occurs both with Ca (ugrandite garnets) and without Ca (pyralspite garnets), and with various amounts of Mg, Fe, Mn, or Al (Deer et al. 1966). Uvarovite, an ugrandite garnet, contains Cr and lacks Al and Fe. Pyrope ($Mg_3Al_2Si_3O_{12}$) is a common garnet in peridotite. Spinel is represented by a series of oxide minerals from $MgAl_2O_4$ to $FeAl_2O_4$, with possible substitutions of several other cations. Chromite ($FeCr_2O_4$) is a common spinel in peridotite and magnetite ($FeFe_2O_4$, or Fe_3O_4) is a common accessory in serpentinite.

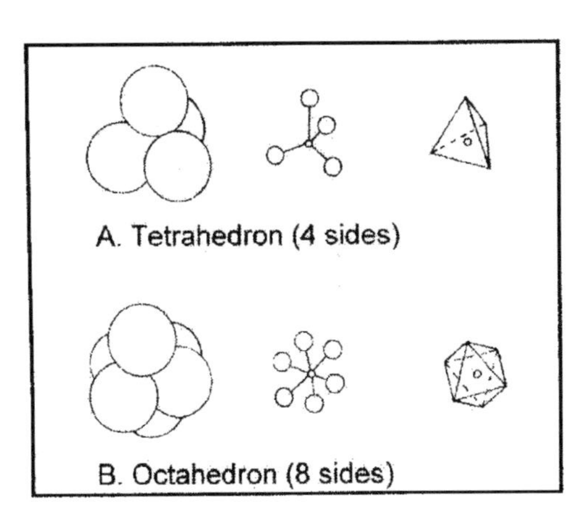

Figure 2-3. Tetrahedral and octahedral polygons

Olivine consists of individual silicon tetrahedra. The tetrahedra are four sided polygons with oxygen ions at the apices and a silicon ion in the center (Fig. 2-3). In olivine there are Mg ions between the tetrahedra. Divalent Fe and other cations of comparable size and valence (Table III.1), such as Ni, can substitute for the Mg in olivine. Pyroxenes are constructed of single chains of silicon-centered tetrahedra (Fig. 2-4). In orthopyroxene, the chains are stacked together side by side to form crystals with three perpendicular axes of symmetry (orthorhombic system). In clinopyroxenes, substituting larger Ca ions for small Mg ions, does not allow for orthogonal stacking of the chains and one crystal axis is no longer perpendicular to the other two, but inclined to them (monoclinic system). Calcium, Na, K, and other alkaline earth and alkali cations larger than Mg that are excluded from olivine and orthorhombic pyroxenes (enstatite) and orthorhombic amphibole (anthophyllite, none of the other orthorhombic amphiboles are common) are called *incompatible* ions. Also, Al with a large ratio of charge to ionic radius is an "incompatible" cation that is excluded from olivine and orthopyroxenes. Most of the Fe in these minerals is divalent.

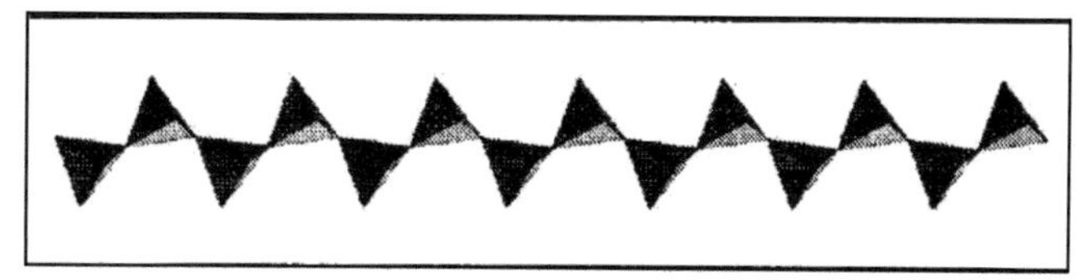

Figure 2-4. A chain of silicon-centered tetrahedra.

Partial melting of lherzolite at the top of the mantle produces mafic magma and leaves a residue with the composition of harzburgite in the upper mantle. Harzburgite is a peridotite with less of the Ca-bearing clinopyroxenes than lherzolite, and no more than traces of the alkali cations. Considerable Cr is retained in harzburgite as a component of chromite.

At oceanic spreading centers, melt (or magma) containing the incompatible ions rises to form gabbro chambers, and magma that rises further upward through dikes to reach the ocean and flow across the ocean floor. The extruded magma cools relatively quickly under water to form pillow basalt. As the magma in gabbro chambers cools, crystals from some of the heavier minerals sink to the bottom and form layers of harzburgite and lenses of dunite (olivine) and pyroxenite (pyroxenes). Layered harzburgite in ocean crust is distinguished stratigraphically from harzburgite that has been deformed, or tectonized, by plastic flow in the mantle (Fig. 1-3).

Plates of oceanic crust consisting of harzburgite (and commonly some lherzolite, dunite, and pyroxenites), gabbro, sheeted dikes of diabase, and pillow basalt drift away from spreading centers until they cool enough to sink down into the mantle in a process called subduction.

Minor amounts of oceanic crust are added to continental crust by various processes, rather than being returned to the mantle. On land, complete sequences of oceanic crust, from peridotite to gabbro and basalt, are called ophiolite (Fig. 1-3). How some oceanic crust has escaped the recycling process and appeared on continental crust, rather than being subducted back into the mantle, has been a mystery to geologists, and some of the mechanisms are still controversial. Wakabayashi and Dilek (2003) have described four possible mechanisms. Upon the addition of water to peridotites in ocean crust or ophiolites, they are readily altered to serpentinite containing serpentine and a few other minerals (section 2F).

C. Volcanic Ultramafic Rocks

Komatiite is the major volcanic ultramafic rock. In North America, it occurs only on the ancient craton that formed during the Archean. It occurs on the ancient cratons of other continents, also, and is named after the Komati River, which is in southern Africa. Picrites are also common Archean volcanic rocks. They have Mg and Fe concentrations between komatiites and basalts (Table 2.2).

Komatiite is defined as having less than 52% SiO_2, more than 18% MgO, TiO_2 <1%, and alkali cations (Na_2O+K_2O) < 2% by weight (Le Maitre 2002). Spinifex texture is a common and distinctive feature of komatiites. This texture is characterized by a network of long acicular olivine crystals that geologists found to be reminiscent of the branching pattern in a grass plant (*Triodia spinifex*) that is common on an ancient craton of western Australia. Komatiite flows require temperatures >1600 C (Arndt 2008). Few komatiites postdate the Archean and komatiite flows no longer occur on Earth. They are widely distributed across the Canadian Shield and small, geoecologically insignificant, bodies occur south as far as the Wyoming Province of the Archean to early Proterozoic craton. The type locality of komatiite in North America is Pyke Hill (Fig. 2-1 and 2-5) in the Abitibi greenstone belt of the Superior Province.

Table 2.2 Concentrations of elements in ultramafic rocks and basalt.

Rock Type	SiO_2	MgO	FeO	CaO	Al_2O_3	TiO_2	MnO	Na_2O	K_2O	P_2O_5
	g/kg									
Webster.	478	192	120	94	61	6.4	0.9	8.4	4.1	1.6
Lherzol.	430	287	119	54	42	4.2	1.7	5.5	2.5	1.1
Harzburg.	417	347	125	30	25	2.6	1.5	3.1	1.4	1.3
Dunite	404	400	137	11	19	0.9	7.1	2.0	0.8	2.0
Basalt(2)	494	71	113	100	153	19.1	2.0	26.4	12.6	3.0
Komatiite	459	292	108	60	65	3.5	1.9	2.2	0.8	0.3
Lherzol.	437	372	100	33	28	2.5	1.3	3.3	1.4	—
Picrit. liq.	446	143	125	132	122	1.0	2.0	11.0	5.0	—
Kom. liq.	463	274	146	53	45	0.4	1.0	5.0	2.0	—

Websterite, lherzolite, harzburgite, dunite, and basalt data from Le Maitre (1976); komatiite data from Arndt (2008); and lherzolite with 2% melting at 1475 C to yield picrite melt and 60% melting at 1700 C to yield komatiite melt (Mysen and Kushiro 1977).

Picritic basalts, or picrites, as well as komatiites, occur in Archean greenstone belts. The proportions of komatiites and picritic basalts decreased through the Archean and early Proterozoic as volcanic rocks became less mafic (Condie 2005). Some later komatiites that are along the trailing western edge of the Caribbean plate (Kerr et al. 1996, Alvarado et al. 1997) may be related to the Galapagos hot spot (Chapter 12). Otherwise no komatiites have been produced in the last billion years.

D. Layered Bodies

Some layered bodies, or intrusions, are in sills such as those in the Stillwater complex of the Wyoming Province. Some layered intrusions such as the Muskox complex in the Bear Province (Fig. 2-1) and the Great Dyke in Zimbabwe are wedge-shaped bodies in dikes, and some such as the Bushveld complex in South Africa and the Skaergaard complex in Greenland are funnel-shaped bodies that are not especially elongated. The geologic properties of these bodies are well-known (Best 2003), because they are commercial sources for nickel, cobalt, chromium, copper, palladium, platinum, and gold.

All of the layered bodies mentioned above, except the 56 Ma Skaergaard complex, are Archean or Proterozoic in age. Layered intrusions of the 1.1 Ga Duluth complex of the Superior Province are mainly gabbro. Other ancient layered complexes all have some ultramafic rocks and some have large amounts of them. The 2.0 Ga Bushveld complex is much larger than the others and has very large amounts of ultramafic rocks. The 1.2 Ga Muskox dike complex (Fig. 2-1) is approximately 120 km long, and up to several km wide (Irvine and Smith 1967), and the Great Dyke of Zimbabwe is about 550 km long. The Stillwater sills cover about 440 km^2 and together they are up to 7 km thick (McCallum 1996).

Layers in the intrusive bodies are differentiated by combinations of crystal fractionation and gravitational separation and by multiple injections of magma. The Stillwater complex has layers of gabbro, norite, troctolite, and anorthosite, which are mafic rocks, and dunite, harzburgite, and bronzite (a pyroxenite), which are ultramafic rocks.

E. Alaskan-type Intrusions

Alaskan-type intrusions are concentrically zoned bodies that are one to several km in diameter. They have a core of dunite and typically grade outward through wehrlite, clinopyroxenite, and hornblendite to gabbro (Himmelberg and Loney 1995). There are several Alaskan-type intrusive bodies in Southeast Alaska and more in British Columbia (Nixon et al. 1997 and sections 8D2 and 8E2 in Chapter 8, Northwest).

F. Metamorphic Ultramafic Rocks, Serpentinite

The metamorphism of peridotite by hydration produces serpentinite. This occurs in ocean crust and continues after peridotite has been placed in continental crust. Most peridotites in continental crust are at least partially serpentinized, but pyroxenites commonly elude serpentinization. Complete alteration of peridotite (density about 3.2-3.3 Mg/m^3) to serpentinite (density ~2.6 Mg/m^3) by the addition of 12 to 14% water increases the volume by about 27%. This increase in volume, or expansion, in a confined space below ground causes fracturing and shearing that produce the smooth surfaces on serpentinite as seen in Figure 1-1. If sufficient magnesia (MgO) and silica (SiO_2) are removed in the alteration of peridotite to serpentinite, there is no expansion and the serpentinite is massive (Fig. 2-5). Serpentinite is easy to identify, without magnification, when it has the smooth surfaces developed from shearing and more difficult to identify when it is massive.

Figure 2-5. Peridotite that was in the process of serpentinization when it was exhumed and exposed to the atmosphere. The exposed peridotite has been weathered and iron from it has been oxidized to produce dark reddish brown oxides.

Serpentinite is mostly serpentine, commonly with some brucite, and with inclusions of magnetite. There are three common varieties of serpentine: antigorite, lizardite, and chrysotile (Appendix III). Lizardite and chrysotile are stable below 300 Celsius (O'Hanley 1996) and antigorite, which is less common, is produced at higher temperatures. Lizardite is a platy mineral and chrysotile is commonly fibrous, as in asbestos. Both minerals have the same chemical composition: $Mg_3Si_2O_5(OH)_4$. Lizardite is the most common mineral in serpentinite and chrysotile is common in veins in serpentinized peridotite and serpentinite (Fig. 2-6). Varieties of chrysotile and lizardite with Ni substituted for Mg are pecoraite and népouite (O'Hanley 1996).

Figure 2-6. Serpentinized peridotite with a vein of serpentine, chrysotile. Protrusion of the chrysotile beyond the weathered surface of the rock indicates the chrysotile is more resistant to weathering than the olivine and the pyroxenes in the peridotite.

Examples of the hydrothermal alteration of olivine and pyroxenes. The serpentinization of olivine containing some Fe (commonly about 1 mole of Fe to 9 moles of Mg) produces serpentine ($Mg_3Si_2O_5(OH)_4$), brucite ($Mg(OH)_2$), and magnetite ($FeFe_2O_4$). Only chromite ($FeCr_2O_4$) may be preserved from the complete serpentinization of peridotite. An example of the alteration of forsterite, which is the Mg end member of olivine, without Fe for simplification, is

$$2\ Mg_2SiO_4 + 3\ H_2O = Mg_3Si_2O_5(OH)_4 + Mg(OH)_2$$

olivine (forsterite) + water = serpentine + brucite

With the possible loss of some Mg and Si during serpentinization, the amount of brucite is reduced, minimizing the increase in volume from peridotite to serpentinite. The Fe from olivine, which is generally much less than Mg in it, is reposited in brucite at a molar ratio about 1Fe:3Mg, and to a lesser concentration in serpentine, and the rest forms magnetite. The distribution of Fe among the minerals formed by the serpentinization of dunite, harzburgite, and lherzolite was illustrated by Alexander (2004). Generation of hydrogen by the oxidation of ferrous iron in the process of serpentinization, $3Fe^{2+} + 4\ H_2O = 6\ H^+ + H_2 + FeFe_2O_4$, can occur in anaerobic environments. Calcium is lost during the serpentinization of clinopyroxenes, or it can be used to form tremolite, $Ca_2Mg_5Si_8O_{22}(OH)_2$, or actinolite in which Fe is substituted for some of the Mg of tremolite, for example

$$4\ MgSiO_3 + 2\ MgCaSi_2O_6 + 2\ H_2O = Ca_2Mg_5Si_8O_{22}(OH)_2 + Mg(OH)_2$$

orthopyroxene + clinopyroxene (diopside) + water = tremolite + brucite

Tremolite and actinolite are amphiboles, which are double chain minerals. With loss of Ca, chlorite or talc can be produced, although some Al, as from garnet, is required to produce a chlorite that is common in serpentinite, for example

$$MgSiO_3 + MgCaSi_2O_6 + Mg_3Al_2Si_3O_{12} + 11\ H_2O =$$
$$MgAl(Si_3AlO_{10})Mg_4(OH)_8 + Ca(OH)_2 + 3H_4SiO_4$$

orthopyroxene + clinopyroxene + garnet + water =
chlorite + Ca-hydroxide + silicic acid

where the orthopyroxene is enstatite, the clinopyroxene is diopside, the garnet is pyrope, and the chlorite is clinochlore. The H_4SiO_4 and $Ca(OH)_2$ from this reaction can be carried into adjacent gabbro and associated rocks and alter the rocks to roddingite, which is not an ultramafic rock. Aluminum is not necessary for the production of talc, for example,

$$2\ MgSiO_3 + MgCaSi_2O_6 + 2\ H_2O = Mg_3Si_4O_{10}(OH)_2 + Ca(OH)_2$$

orthopyroxene + clinopyroxene + water = talc + Ca-hydroxide

Rocks composed of talc are called soapstone, or steatite, and the formation of soapstone from pyroxenes is called steatization. Soapstone is an ultramafic rock that is easy to identify, because it is soft (scratched by a finger nail) and feels smooth. It is much less common than serpentinite, and it can be produced from the alteration of dolomite or marble with a source of silica, as well as from ultramafic rocks. Serpentinization of olivine with CO_2 present produces magnesite ($MgCO_3$,

Fig. 2-7), along with serpentine.

$$2\ Mg_2SiO_4 + 2\ H_2O + CO_2 = Mg_3Si_2O_5(OH)_4 + MgCO_3$$

olivine + water + carbon dioxide = serpentine + magnesite

G. Hydrology

There are few dominantly ultramafic watersheds in North America that are large enough to have second or third order streams, which are streams that are fed by first order streams from the upper reaches of a watershed. Two ultramafic watersheds in North America with second and third order streams have been monitored in flumes as stream water has flowed from the watersheds. These two watersheds are the 3650 ha Clear Creek drainage in California (Alexander et al. 2007a) and a 57 ha watershed at Soldiers Delight in Maryland (Cleaves et al. 1974). A few more entire ultramafic watersheds have been monitored in New Caledonia (Trescasas 1975) and in the Czech Republic (Kram et al. 2012).

Figure 2-7. Magnesite veins of small and larger sizes in serpentinized peridotite on the Vizcaino Peninsula, Baja California Sur. Blocks of magnesite have been mined where the veins are thicker than 0.2 meters.

The Clear Creek ultramafic watershed is in a temperate, summer-dry climatic zone of the California Coast Ranges. Practically all of the smaller streams in the Coast Ranges are dry through summers. Stream flow from the small Clear Creek watershed, however, continues longer into the spring than flow from nonultramafic watersheds of comparable size in the Coast Ranges. Stream runoff from Clear Creek peaks in March, after runoff peaks from small nonultramafic watersheds in the California Coast Ranges in February (Alexander et al. 2007a).

Soldiers Delight is in a temperate, humid climatic zone east of the Appalachian Mountains. In the Soldiers Delight area there is ephemeral stream flow through summers from the ultramafic watershed, but not from an adjacent feldspar-quartz-biotite schist watershed (Cleaves et al. 1974).

The reason for delayed flow from the ultramafic watersheds may be extensive fracturing in weak serpentinite bedrock, allowing the storage of extra quantities of groundwater. Magnesium greatly exceeded the concentrations of other cations in water draining from both the Soldiers Delight and Clear Creak watersheds. The pH of drainage water from the ultramafic watersheds averaged 8 at Soldiers Delight and 8.9 from the Clear Creek watershed.

Springs fed by water that flows through peridotite that is currently being serpentinized are very strongly alkaline (pH >11.0). These springs occur in many places around the world (Barnes et al. 1967, 1978). By measuring the hydrogen (deuterium) and oxygen isotope ratios (δ^2H and δ^{18}O), Morrill et al. (2013) determined that the serpentine spring water was meteoric in the Austin Creek watershed of the California Coast Ranges. Water from all of the serpentine springs sampled along Austin Creek had much higher Ca than Mg concentrations and some had very high Na and Cl concentrations. They identified the former as waters that circulated predominantly through peridotite and the latter as water that circulated to greater depth and reacted with both peridotite and nonserpentine rocks of the Franciscan complex.

Rain water that enters the soils above Austin Creek is moderately acid. After passing through the soils and reacting with Ca bearing pyroxenites in peridotite beneath the soils, the water returns to the surface in springs from the ultramafic strata. The water in the springs is concentrated with Ca-hydroxide, and with both $Ca(OH)_2$ and NaOH where the water has passed through both ultramafic rocks and nonultramafic rocks of the Franciscan complex. Once this very highly alkaline spring water is exposed to CO_2 of the atmosphere, the calcium is precipitated as a carbonate ($CaCO_3$), or travertine, around the springs or just below them. The travertine can form levees around springs to create pools of spring water (Fig 2-8). Short distances downstream from springs of the ultramafic strata, following the precipitation of Ca-carbonate, Mg^{2+} and bicarbonate become dominant ions in the stream water, and the pH of the stream water declines. The average pH of water in Austin Creek, downstream from the serpentine springs, was ascertained to be 8.7 (Morrill et al. 2013).

Stream water has been collected several places along Rough and Ready Creek (Miller et al 1998), which drains a dominantly ultramafic watershed in the Klamath Mountains of western North America (Alexander et al. 2007b). The stream water was slightly alkaline, ranging from pH 7.6 to 8.5. It had high Mg and low Ca concentrations. The concentrations of silica and bicarbonate were high relative to average stream waters of the world, and the water had exceptionally high Ni and high Cr concentrations.

Many streams draining from ultramafic terrain in the Klamath Mountains have stream beds that are cemented with silica (Alexander 1995). Good, easily accessible examples of the silica cemented stream beds are along High Camp Creek near the head of the Trinity River, and in Hayfork Creek where the Wildmad Road runs along the west side of the creek.

Figure 2-8. A travertine pool along a tributary of Austin Creek.

3 Weathering and Soils

Ultramafic soil forming processes are mostly the same as those in other soils. A major difference is that the lack of aluminum (Al) in ultramafic soils inhibits the production of aluminous clay minerals. Although some of the physical properties of peridotite soils differ from those of serpentinite soils the properties of ultramafic soils are not unique. They are like those in many other kinds of soils, other than those with large amounts of amorphous Al-silicates.

Most ultramafic soils develop over peridotite or serpentinite bedrock or in colluvium from ultramafic strata. There are few large areas of ultramafic rocks where fluvial deposits are mostly from ultramafic sources. Generally alluvium in areas with ultramafic rocks is mixed with enough nonultramafic alluvium that the soils are not considered to be ultramafic soils. There are few exceptions. Soils in ultramafic fluvial deposits are featured in sections 9C3 and 10D. Soils in ultramafic glacial till are in section 9C4.

It is the chemical compositions of the minerals in the ultramafic rocks of soil parent materials that are unique. Although the physical properties of ultramafic soils are not unique, it is important to discuss them. There are many misconceptions about ultramafic soils.

A. Soils and Weathering

Most of the activities of plant roots and many microorganisms are in soils. The *in situ* production of fine-earth (particles <2 mm) from ultramafic rocks, without abrasion in flowing water, ice, or wind, is by *weathering* (defined in the Glossary). Freeze and thaw may be appreciable in the weathering of some very cold soils. Otherwise, most of the weathering in ultramafic soils is by chemical processes, and living organisms are involved in much of that weathering. Much of the weathering of ultramafic rocks, or rock fragments, occurs at the bedrock-soil interface and within soils. The bedrock-soil boundary is commonly indistinct and irregular (Fig. 3-1).

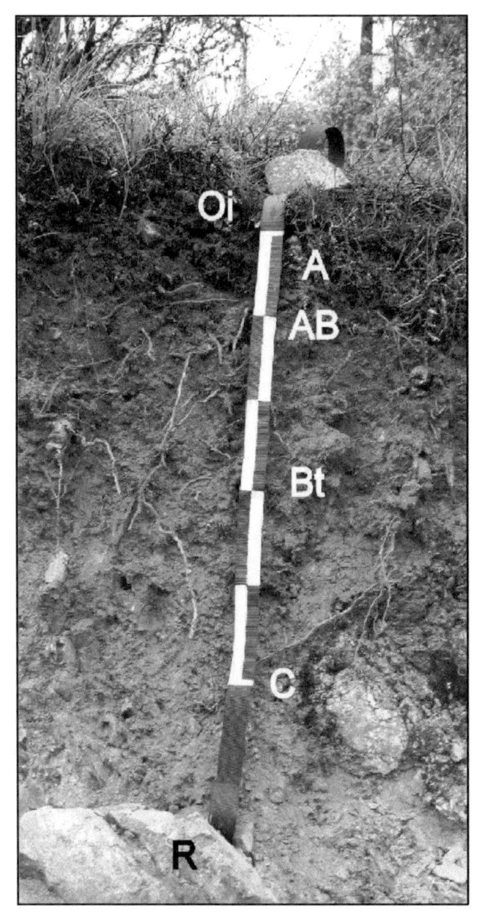

Figure 3-1. A typical ultramafic soil in the Klamath Mountains, a moderately deep Alfisol It is in an open forest with understory shrubs and grasses.

Ultramafic rocks consist primarily of Si and Mg, with substantial Fe and lesser amounts of Ca and Al. The Si, Mg, and minor Al, form clay minerals and in well drained soils, which are the great majority of ultramafic soils, the Fe is oxidized to impart reddish colors to the soils (Fig. 3-1). Some Fe is commonly concentrated in nodules that upon drying become hard. Pea-size black nodules in ultramafic surface soils have been called "iron shot" by miners (Ramp 1978). I have found concentrations of them on the eroded surface of an old Gasquet soil, an Ultisol (Kandihumult). They were practically all magnetic, implicating magnetite, or possibly maghemite, as Matsusaka et al. (1965) found in soils developed from basalt and andesite.

Chemical weathering of ultramafic rocks and weathering in ultramafic soils is generally aerobic relative to serpentinization, which is by anaerobic processes. Relatively few ultramafic soils are poorly drained, although there are some (for example, Lee et al. 2003), and some are wet organic soils (Histosols) with ultramafic substrata (Chapter 9, section 9C5).

Weathering processes and rates are different in different climates with different plant communities. Both higher temperatures and greater precipitation contribute to increased weathering. Some details of the weathering and soil development differences are presented in chapters 5 through 12 and in the appendices.

B. Soil Physical Properties

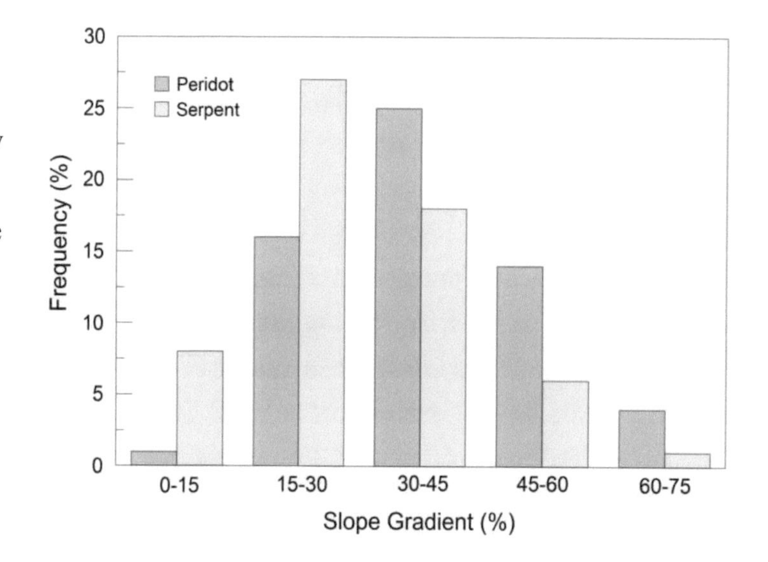

Figure 3-2. Slope gradient distributions on 12,000 hectares of ultramafic (serpentinite and peridotite) soils, excluding large landslides, based on GIS measurement of gradients on 30 m slope lengths at 120 randomly chosen locations each in peridotite and serpentinite soil map units. The GIS data were obtained by Jonna DuShey, McCloud District, Shasta-Trinity National Forest.

Ultramafic soils range from sandy to clayey, and from nonstony to very stony. And from very shallow to very deep. There is a common perception that ultramafic soils are more shallow than those with other parent materials, which is possibly correct, but it would be difficult to prove because of the wide ranges of soil depths on most soil parent materials, including ultramafics. In a detail survey of 12,000 ha of ultramafic soils in steep terrain on the southern exposure of the Rattlesnake Creek terrane in the Klamath Mountains, 196 pedons (74 with peridotite and 122 with serpentinite soil parent materials) were described to represent the soils mapped there (Alexander 2003). Even though slope gradients tend to be greater on peridotite than on serpentinite soils (Fig. 3-2), the depth distributions of peridotite and serpentinite soils (Fig. 3-3) are similar, and they appear to be similar to those of other kinds of soils in steep terrain on the Shasta-Trinity National Forest (Lanspa 1993), which is where the Rattlesnake Creek terrane is located. Peridotite is generally more massive than serpentinite, which is commonly sheared by tectonic forces. The tectonic shearing produces smooth surfaces that are characteristic of serpentinite in accreted terranes. This can contribute to diminished slope stability. Also, peridotite soils are generally more stony than the serpentinite soils (Alexander and DuShey 2011). The very deep peridotite soils in the Rattlesnake Creek terrane are mostly in colluvium down steep and very steep slopes and the very deep serpentinite soils are generally on landslides with slope gradients <30% (Alexander 2003).

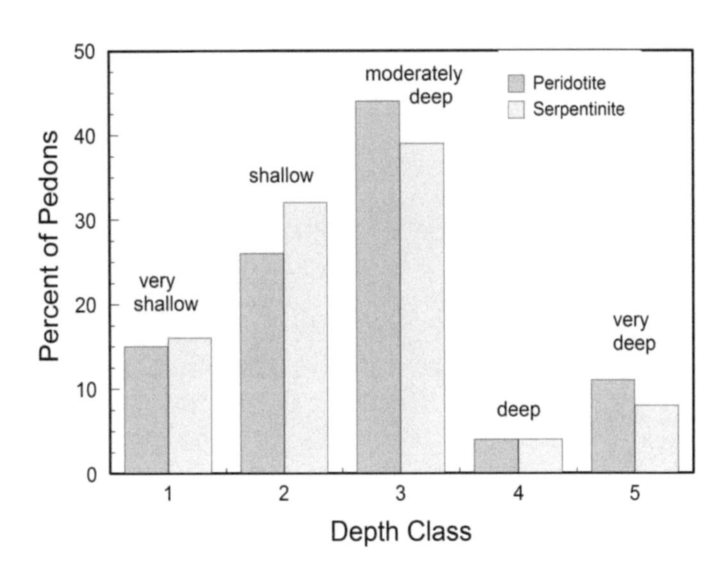

Figure 3-3. Soil depth distributions of 74 peridotite and 122 serpentinite pedons described to represent the soils of the southern exposure of the Rattlesnake Creek terrane in the Klamath Mountains of western North America. Depth classes: (1) 10-25 cm, (2) 25-50 cm, (3) 50-100 cm, (4) 100-150 cm, and (5) depth >150 cm.

Weathering and soil depths are related to precipitation and soil moisture regimes, among the other things. At, or near, the 40°N latitude of the Rattlesnake Creek terrane, moderately deep Alfisols are representative of ultramafic soils in cool, open forests and shallow Mollisol are representative of ultramafic soils in warm grassland and chaparral that are drier.

Water storage capacities of ultramafic soils in the Klamath Mountain are similar to those of other kinds of soils in mountainous terrain there (Alexander et al. 2007a), except where there is appreciable volcanic ash. Soils developed from volcanic ash generally hold more water. There is no reason to believe that ultramafic soils characteristically have lower water storage capacities than other soils, other than those with volcanic ash. The xerophytic characteristics of many serpentine plants is commonly cited as evidence that ultramafic soils are more droughty than others, but the xerophytic characteristics of plants on ultramafic soils may be related more to the unusual chemistry of ultramafic soils than to droughtiness.

C. Soil Mineralogy

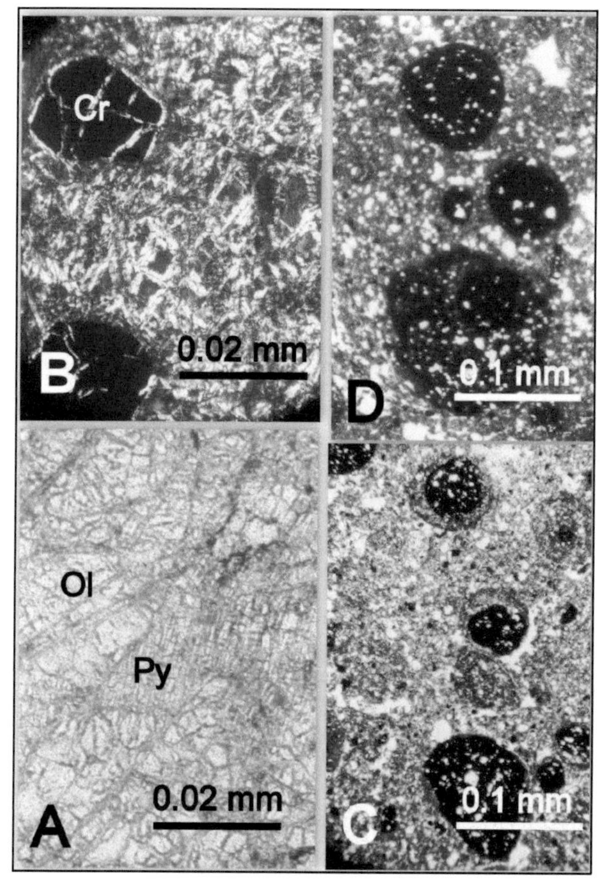

Figure 3-4. Weathering in ultramafic soils and parent materials of Ultisols in the Klamath Mountains. A. Fracturing and cleavage of fresh olivine (Ol) and pyroxene (Py). B. Serpentinization and rock weathering, with sparse remnants of pyroxene (blue interference color), serpentine (white), and chromite or magnetite (black). C. A lower B horizon, and D. A redder upper B horizon. Nodules have formed in the B horizons; white spaces are voids and the matrix in the upper B horizon is stained red with Fe oxides.

The weatherable minerals in partially serpentinized peridotite are olivine, pyroxenes, serpentine, and brucite (Fig. 3-4). Brucite occurs with serpentine in serpentinite and does not resist weathering long enough to become a noticeable constituent in soils. Olivine is readily weatherable, with the release of considerable Mg, Si, and Fe and the formation of Mg-smectite (saponite) and goethite (Velde and Meunier 2008). Pyroxenes can weather to produce vermiculite and a variety of other clay minerals. Serpentine is more resistant to weathering than olivine and pyroxenes and can persist in soils for long periods.

Serpentine, chlorite, and magnetite in ultramafic soils are inherited from serpentinized rocks. The most common secondary clay minerals in ultramafic soils are smectite and goethite, and in arid soils palygorskite (Appendix III). Chlorite from the weathering of pyroxenes (Caillaud et al. 2006) or released from the weathering of serpentinite can weather via interstratified chlorite/vermiculite to vermiculite (Rabenhorst et al. 1982, Lee et al. 2003). With time, smectite commonly becomes the dominant soil clay mineral, except in arid areas where palygorskite might become dominant (Alexander 2019), and in wet climates where goethite might become dominant (Alexander 2010). Vermiculite can be a common mineral in leached soils of wet climates (Paquet and Millot 1972, Burt et al. 2001). The dominant clay mineral in very old, leached ultramafic soils in humid climates is goethite, and there is generally enough hematite to make the old soils red with 2.5YR and 10R hues (Alexander 2010). Kaolinite can occur in weathered and leached ultramafic tropical soils (Appendix IV).

D. Soil Chemistry and Plant Nutrients

Most of the clay minerals in soils have negative charges, which are cation exchange sites that can be occupied by Mg, Ca, Na, K, and H ions. Some exceptions are old, moderately acid ultramafic soils dominated by goethite and other Fe oxides that have more positive than negative charges (Alexander 2010, 2014a). Alkali (monovalent) and alkaline earth (divalent) cations are easily replaced by one another from the cation exchanges sites and they are readily available to plants. Because the exchangeable Ca:Mg ratios are generally low (molar Ca/Mg <0.7), or much lower below the surface horizons in ultramafic soils, plants may not get enough Ca for optimum growth. Also, exchangeable K contents are generally lower in ultramafic soils than in other soils. Additional problems for plants are low phosphorus (P) and molybdenum (Mo) concentrations in ultramafic soils. These deficits limit the growth and survival of plants on ultramafic soils, but they are advantageous for the survival of some plants that cannot compete with other plants on many soils but can cope better than other plants with the limitations of ultramafic soils.

Soil organic matter contents in ultramafic soils are comparable to those in nonserpentiine soils, and the C:N quotients (C/N ratios) are similar (Alexander et al. 2007a). Therefore, there is no reason to believe that the range of N contents in ultramafic soils is less than in most other soils. Some exceptions are soils with parent materials of clayey sediments where N-bearing relics of once living microorganisms can be preserved for millions of years, or soils developed from NH_4-bearing schists derived from those sedimentary rocks by metamorphosis at temperatures >500 C. Thus, substantial soil N can be derived from the weathering of some soil parent materials (Dahlgren 1994), and much of the extra N can be available to plants (Morford et al. 2016). Nevertheless, N is no more a plant growth limiting factor in ultramafic than in many nonultramafic soils; N is a limiting factor in most soils.

Ultramafic soils are commonly slightly to moderately acid or neutral, even in some old, leached soils (Alexander 2010). Ultramafic desert soils sampled in Baja California are neutral in the surface to slightly or moderately alkaline in subsoils (Chapter 10, 10D). Few ultramafic soils have any horizons that are calcareous, and none of them have calcic horizons.

Some of the first transition elements (defined in Appendix II) from Cr through Fe to Ni have relatively high concentrations in ultramafic soils and may be toxic to plants. Nickel is the most likely element to be toxic to plants. Although Ni appears to be toxic to some plants in ultramafic soils, it is generally difficult to differentiate the plant growth and survival limitations of Ni toxicity from limitations caused by low Ca/Mg concentrations. Plants that are highly susceptible to Ni toxicity do not grow on ultramafic soils. Concentrations of first transition elements Sc and Ti, or V, to Ni, and possibly Cu and Zn, are all relatively high in mafic rocks, compared to more silicic rocks (Turekian and Wedepohl 1961). All of the elements from V to Ni accumulate in soils that are weathered and leached. Manganese and Fe are generally more concentrated in ultramafic soils and Cr, Co, and Ni are very much more concentrated in ultramafic soils than in other soils. The very old Littlered soil, an Ultisol (Kandihumult) of limited productivity, has 1.3% total Ni in the subsoil, which could be a factor in its low productivity (Alexander 2010).

With only minor amounts of nonsepentine materials in the parent materials, old ultramafic soils become more productive as they age, while old soils with strictly ultramafic parent materials are not very productive. The differences can be observed by comparing plant communities on old ultramafic soils in California (Alexander 2010). Figure 3-5A shows Walnett soil with only serpentinized peridotite parent material, no nonultramafic materials (Fig. 3-5A), and Forbes soil with barely detectable remnants from andesitic lahar that covered the serpentinized peridotite parent material millions of years ago (Fig. 3-5B). Another comparison to the Walnett soil (Fig. 3-5A, mean annual precipitation, MAP=3000 mm) is the Brockman variant (another Haploxeralf, MAP=1500 mm), which is in predominantly serpentinized peridotite outwash, but with 19% nonultramafic gravel, mostly gabbro. The Brockman variant is much more productive than the Walnett soil. Based on tree growth (Chapter 9, 9C3), the Brockman variant is highly productive. Physically the soils are similar, both are Alfisols (Haploxeralfs) and there is no obvious reason based on the laboratory chemical data (Alexander 2010) for the Brockman variant being much more productive than the Walnett soil. Although the exchangeable Ca concentrations are similar in both soils, the Ca reserves, based on the total analyses, indicate there is much more Ca in the Brockman variant (Alexander 2007b); and the amounts of K and P, although low, are greater in the Brockman variant than in the Walnett soil. Thus, minor differences in parent materials can result in old ultramafic soils with similar parent materials having major differences in forest productivity.

Figure 3-5. Two similar old ultramafic soils in western North America, with much different vegetation. A. A forest of low productivity on the deep Walnett, an Alfisol (Haploxeralf) in the Klamath Mountains. B. A highly productive forest on the Forbes, an Ultisol (Kandihumult) in the Motherlode sector of the Sierra Nevada.

E. Kinds of Ultramafic Soils

Most ultramafic soils have surface A and subsurface B horizons and those in the colder or wetter climates generally have surface organic (O) horizons, which are designated Oi, Oe, or Oa horizons (Soil Survey Staff 1999, 2014). The B horizons can be cambic (Bw), argillic (Bt), silicic (Bq), silicic and indurated (Bqm), oxic (Bo), or with slickensides developed in clayey soils by shrink and swell during wetting and drying (Bss horizons). Very few ultramafic soils have any calcareous (Bk) horizons and none in North America have been found to have calcic horizons. Old, highly weathered and leached ultramafic soils commonly have kandic B horizons.

Table 3.1 The twelve orders of Soil Taxonomy.

Order	Defining Properties
Alfisols	argillic, kandic, or natric horizons
Andisols	amorphous Al (or Fe) compounds and low density, with or without volcanic glass
Aridisols	dry most of the time, with argillic, cambic, calcic, or other diagnostic horizons
Entisols	no diagnostic horizons, rudimentary soils
Gelisols	frozen soils
Histisols	organic soils
Inceptisols	cambic horizons, inception of soil development
Mollisols	mollic epipedons, with or without argillic horizons
Qxisols	oxic horizons, extremely weathered soils
Spodisols	spodic horizons, accumulation of translocated Al or Fe
Ultisols	argillic horizons with low basic cation status
Vertisols	cracking-clay soils, vertical mixing of soil

Ultramafic soils occur in at least ten of the twelve orders of Soil Taxonomy (Soil Survey Staff 1999): Alfisols, Aridisols, Entisols, Histosols, Inceptisols, Gelisols, Mollisols, Oxisols, Ultisols, and Vertisols (Table 3.1). Spodosols can occur where ultramafic materials are mixed with or covered by nonultramafic materials (Alexander et al. 1994a, D'Amico et al. 2008). The majority of ultramafic soils in North America are in three orders: Inceptisols, Mollisols, and Alfisols. Aridisols may be common in ultramafic areas of deserts, and Oxisols may be common in tropical areas with ultramafic rocks. Ultramafic Gelisols are found far north in the Ural Mountains, but none have been reported in North America.

The more common kinds of ultramafic soils in North America are differentiated primarily by the thicknesses of dark colored, surface, A, horizons and the properties of subsoil, B, horizons (Fig. 3-6). The rudimentary Entisols lacking B horizons and the highly weathered Oxisols, might be considered the end members of a soil development sequence. The ages and B horizon development of the Mollisols with Bw (cambic) horizons are similar to those of Inceptisols, and the ages and B horizon development ranges of Mollisols with Bt (argillic) horizons are similar to those of Alfisols. The Mollisols are distinguish by having thick, dark colored A horizons. Ultisols are leached of basic cations (Ca, Mg, Na. K) such that the basic cations are no longer dominant over acidic cations (H and Al) on the cation exchange complexes. Ultisols are not necessarily older than Alfisol and Mollisols, but in North America they were first recognized in nonglaciated areas east of the Appalachian Mountains. There are no Ultisols in those glaciated areas east of the Appalachian Mountains, which are less than 125 ka.. Along the Pacific Coast, Ultisols are found in areas of high precipitation where the soils are intensively leached , not necessarily in old soils. Oxisols with ultramafic parent materials are not present north of the Caribbean area, except on the Klamath Mountain peneplain.

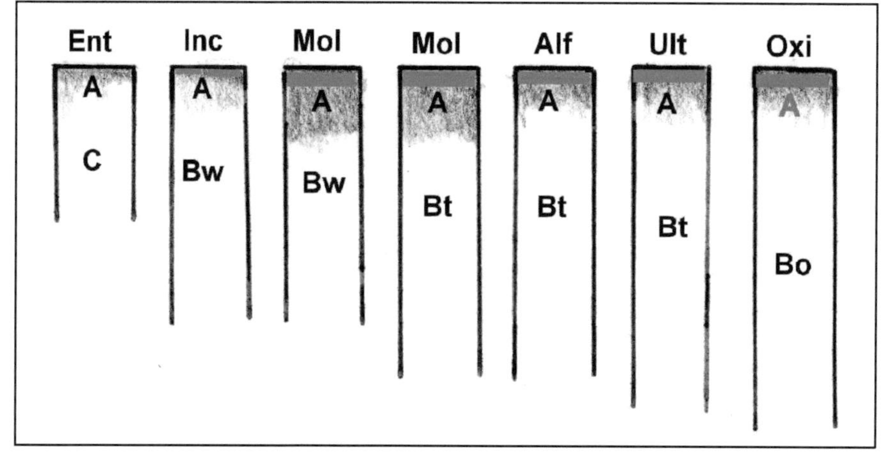

Figure 3-6. Kinds of ultramafic soils representative of those in North America and the Caribbean area: Entisols, Inceptisols, Mollisols, Alfisols, Ultisols and Oxisols.

The terminology of soil classification is simplified for readers who are not familiar with soil taxonomy by applying vernacular modifiers (Table 3.2) to the Orders and relegating more complete taxonomic names to parentheses; for example, argillic Mollisols for Argixerolls or Argiudolls, and haplic Mollisols for Haploxerolls or Hapludolls. Frozen, very cold, cold, cool, warm, and hot are substituted for gelic, cryic, frigid, mesic, thermic, and hyperthermic soil temperature regimes (STRs).

F. Soil Parent Material, Peridotite or Serpentinite

Some ultramafic soil parent materials are definitely peridotite and some are definitely serpentinite, but many are more indefinite. Quick (1981) found that most of the peridotite in the Trinity body of the Klamath Mountains was about 20 to 50% serpentinized. Only in small areas did he find peridotite that was less than 10% serpentinized, with serpentinization generally most intense in fault zones and adjacent to diabase dikes and other intrusive bodies. Other than in Alaskan-type intrusion of Alaska and Canada, the only large bodies of nonserpentinized ultramafic rock I have found in western North America were the core of the dunite body in the Twin Sisters of the Northern Cascade mountains and the wherlite of Monumental Ridge in the Sierra Nevada.

Where unweathered bedrock can be found beneath an ultramafic soil, the specific gravity (SG) is the easiest way to judge the degree of serpentinization. The density (or SG) of peridotite soil parent material is commonly nearer the median between a 2.6 of serpentinite and the 3.2 (or 3.3) of unserpentinized peridotite than near 3.2 kg/m^3.

An alternative index of serpentinization for soils that are not too weathered is the proportion of light minerals (SG<2.89) in a separation of fine sand grains in bromoform, following the removal of *free* iron (Appendix V). The light minerals in soils from ultramafic rocks are almost exclusively serpentine and commonly some chlorite. These are the main minerals in serpentinite.

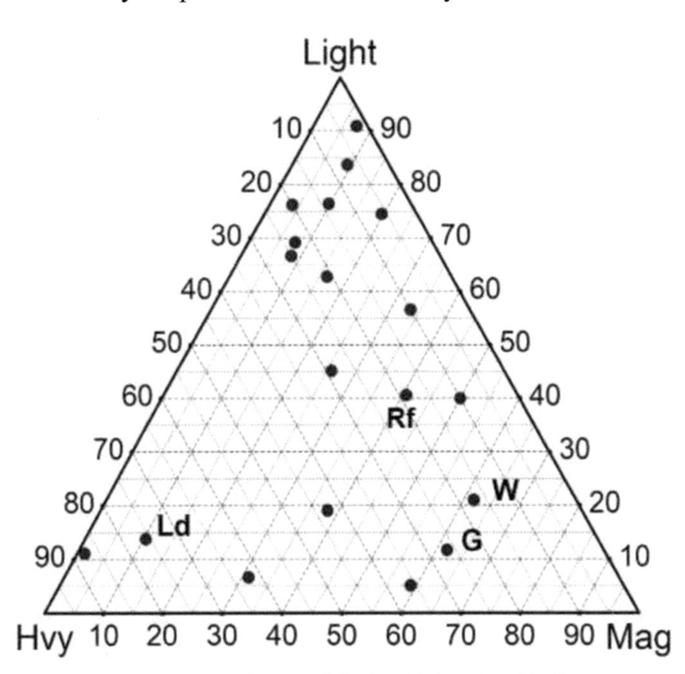

Figure 3-7. Proportions of light (SG <2.89), heavy, and magnetic grains in fine sand fractions of ultramafic subsoils in a transect across the Klamath Mountains at about 42 degrees N latitude. Ld, Low Divide; Rf, Red Flat; G, Gasquet soil; W, Walnett soil.

The heavy minerals in ultramafic rocks that most commonly have some fine sand grains in soils are olivine, pyroxenes, amphiboles, chromite, and magnetite. Magnetite is the only magnetic mineral that is common in most soils, and maghemite is a magnetic mineral that is common in some soils.

Light, heavy, and magnetic fine sand grains were separated from the B horizons samples of 15 moderately deep soils and a deep soil at Red Flat (Curry County, OR) in a transect across the Klamath Mountains (Alexander 2014). Five of the 7 soils, with light fractions >65%, were judged to be serpentinite soils when they were sampled, and none of the 8 moderately deep soils with light fractions <65% were judged to be serpentinite soils. This correspondence between field judgement and evidence from laboratory measurement of fine sands confirms that the fine sand composition of the moderately weathered, moderately deep soils is a good index for separation of peridotite and serpentinite soils. It did not appear to matter that the soils contained more silt plus clay than sand.

A shallow ultramafic soil at Low Divide in Del Norte County, CA, was sampled where the dunite, or wherlite, there was less completely serpentinized than the peridotite in most other ultramafic rocks in the Klamath Mountains. A subsoil sample had 14% and the surface soil had 16% light grains in the fine sand, indicating only slight serpentinization of the soil parent material.

In contrast to the Low Divide soil, the highly weathered Walnett soil and very highly weathered Gasquet soil (Alexander 2010) had greatly declining amounts of light minerals from the subsoils to the surface soils. The light minerals in the fine sand fractions of the bottom (below a meter), subsoil (B horizons), and surface horizons were 77, 21, and 3% in the Walnett soil and 50, 11, and 0% in the Gasquet soil. The heavy fine sand grains in the Walnett and Gasquet surface soils were black, magnetic grains, yellow-brown to dark reddish-brown weathered, unidentifiable mineral grains, and reddish-brown nodules that had been concentrated by the loss of the more weatherable minerals. The Walnett and Gasquet samples are near the magnetic corner of the fine sand grain triangle (Figure 3-7), a corner representing strongly weathered soils with sand grains that are resistant to weathering. Magnetite, and possibly maghemite, are the magnetic minerals magnetic minerals in these soils.

Two moderately deep soils were sampled on Red Hill in the Feather River belt to show differences between soils with peridotite and with serpentinite parent materials. The light fractions of fine sands in the Bt horizons, at about 30 to 48 cm depths, were between 50 and 59% in the soil with serpentinized peridotite and >75% in the soil with serpentinite parent material (Table 9.1). The two soils were distinctly different, both morphologically and chemically.

The fine sand composition in ultramafic soils of the Blue Mountains was more perplexing. Light grains accounted for more than 70% of the fine sand in the subsurface and subsoil horizons of all seven soils sampled in the Strawberry Range and Aldrich Mountains (Table 9.9), even though some of the soils were near outcrops of ultramafic rocks that were no more than slightly serpentinized. Ultramafic soils sampled on a ridge northwest from Vinegar Hill had substantial quartz, feldspars, and biotite, which may have been from nonultramafic rocks that were closely associated with the ultramafic rocks, and the soils had some volcanic glass in them. The volcanic glass is expected to be from Mount Mazama, which was about 320 kn southwest of Vinegar Hill. That raises the possibility that some of the serpentine in ultramafic soils of the Strawberry Range might have been aeolian, but no more than minor amounts of light mineral grains other than serpentine and chlorite were found in the fine sands. That minimizes the likelihood of any major aeolian input.

Table 3.2 Modifiers of soil order designations.

Modifier	Description (not necessarily complete definitions)
alkaline	soil pH >7.2
andic	andic soil properties in horizons insufficiently thick for an Andisol
argillic	with an argillic horizon of illuvial clay accumulation
aridic	dry more than half of the time that the soil is warm enough for plant growth
calcareous	free Ca-carbonate, ultramafic soils can be calcareous but lack calcic horizons
cambic	with a cambic horizon of soil alteration, lacking an argillic horizon
cobbly (texture)	rock fragments >35% by volume, predominantly cobbles 75-250 mm
cold	soils with mean temperatures <8°C (frigid soil temperature regime, STR)
cold, very	soil temperatures <8°C that remain cold through summers (cryic STR)
cool	mean annual soil temperatures between 8 and 15° C (mesic STR)
duric	with a duripan, a hard layer cemented by silica, Bqm and Cqm horizons
fibric	organic soil material with at least 75% fiber after rubbing
gravelly (texture)	rock fragments >35% by volume, predominantly pebbles 2 to 75 mm
haplic	simple, no special features, other than those required for the soil order or suborder
hemic	organic soil with intermediate fiber content, between fibric and sapric
histic	surface organic layer between 20 and 40 cm thick, or 60 cm if fibrous
hot	soils with mean temperatures between 22 and 30°C (hyperthermic STR)
humic	thick surface soil (generally A horizon) rich in organic matter
kandic	with subsurface accumulation of low activity clay (CEC<160 mmol+/kg)
lithic	hard rock within 50 cm depth
melanic	thick, black surface soil with organic C>6%
moist (soil)	soil not dry in any part for more than 90 days in a year
mollic	with a mollic epipedon, a thick, dark, nearly neutral (pH) or alkaline surface horizon
poorly drained	wet or ephemerally wet soil, redoximorphic features within 50 cm
	somewhat poorly drained soils, redoximorphic features within 75 cm
rhodic	dark reddish brown or red, 2.5YR or 10R hue, value moist <3 and dry <4
rocky	rock outcrop >5% of ground surface area
saline	electrical conductivity (EC) of saturated soil extract >4 mmhos/cm (>40 S/m)
salty	containing salts that are more soluble than gypsum
sandy	sand, loamy sand, or sandy loam, and coarse fragments <35% by volume
sapric	organic soil material with no more than about 15% fiber after rubbing
shallow	depth <50 cm to a root restrictive layer, lithic if depth to bedrock <50 cm
shallow, very	depth <25 cm to a root restrictive layer
silicic	some silica cementation, but less than a duripan
skeletal	65-90 % volume of gravel, or of 4-250 mm rock fragments, in a soil
	subskeletal, 35-65 % volume of gravel, or of 4-250 mm rock fragments
stony (surface)	coarse fragments >80 mm in size cover more than 3% of ground surface
stony (texture)	rock fragments >35%, predominantly stones 250-600 mm
terric	a Histosol with mineral (inorganic) substrata within about 90 cm depth
warm	soils with mean temperatures between 15 and 22°C (thermic STR)
wet (climatically)	soil continuously moist in all horizons
wet (undrained)	high water-table

Terms for soil temperature are based on mean annual temperatures at 50 cm depths.

4 Vegetation and Plant Communities

The many unique plants and distinctive plant distributions on ultramafic soils have attracted considerable attention (Kruckeberg 1984, Brooks 1987, Rajakaruna et al. 2009, Harrison and Rajakaruna 2011), much more attention than the soils. Differences in the sizes and densities of plants on ultramafic and nonultramafic soils are commonly so great that anyone can see that they have vegetation differences. The differences may be described broadly by life forms, or physiognomy, or described taxonomically by the dominant species. Plant communities are conveniently designated by combinations of life form and dominant species. In some areas, the cover of terricolous (soil) lichens may be great enough to include then in plant community designations.

Limitations of ultramafic soils for plant survival and growth, compared to other soils, are largely chemical. General effects of the limitations on plant communities can be recognized in natural landscapes, but controlled experiments, generally with individual species, are necessary to learn more about the specific causes of limitations.

Figure 4-1. Representative serpentine barrens. A. Lassics area in the California Coast Ranges. A lithic Entisol with sparse *Phacelia corymbosa* and traces of *Lupinus constancii* and *Eriogonum nudum*. B. Soldiers Delight on the Piedmont east of the Appalachian Mountains. A moderately deep Alfisol in a pine (*P. virginiana*) savanna with blackjack oak (*Q. marilandica*) and post oak (*Q. stellata*) and many grasses, including little bluestem, threeawn, dropseed, and Indiangrass.

A. Vegetation

The vegetation of a geoecosystem is the combination of all plant life in the system. It is indicative of the kinds and quantities of organisms that can be supported on the land, which are major concerns in land management.

Serpentine vegetative cover ranges from barrens to dense forest, although dense forest is not common on ultramafic soils. The serpentine barrens in western North America are generally bare ground with sparse forbs, whereas the common barrens in eastern North America are pine savanna (Fig. 4-1). Pines have been encroaching on serpentine savannas in the Appalachian Mountain area since burning has been suppressed there (Tyndall 1992). Dense forests on ultramafic soils are more common in eastern North America and in the Caribbean area than in western North America.

In western North America ultramafic barren areas are drastically different from peridotite to serpentinite (Fig 4-2). Extreme dunite and peridotite barrens consist of talus on steep slopes, or below steep slopes (Fig. 4-2A). Large exposures of these kinds of barrens occur around Twin Sisters in the Northern Cascade Mountains and many other places in western North America. Extreme serpentinite barrens are generally in detritus from tectonically sheared rock (Fig. 4-2B). The largest exposures of extreme serpentinite barrens are in the Clear Creek (or New Idria) area of the Diablo Range in the California Coast Ranges (chapter 10).

Some physiographic forms of serpentine vegetation, other than the extremes of barrens and dense forests, are alpine tundra, alpine fell-fields, chaparral, savannas, open canopy forests, sagebrush, sparse desert scrub, and fens with graminoids and forbs (Fig. 1-4). Pitcher plant fens are minor, but distinctive, features of ultramafic landscapes (Fig. 4-3).

Figure 4-2. Extreme ultramafic barrens. A. Dunite and peridotite talus below a steep slope in the Klamath Mountains. B. Serpentinite barrens on very shallow soils in the Lassics area of the California Coast Ranges.

B. Plant Communities

Plant communities are commonly designated by the dominant species of plants in tree, shrub, and forb layers, with the layers separated by slashes (/); for example, Jeffrey pine–incense cedar/buckbrush/Idaho fescue–wheatgrass, or *P. jeffreyi–Calocedrus decurrens/Ceanothus cuneatus/Festuca idahoensis–Pseudoregnaria spicata*. Trees are placed ahead of shrubs and forbs because they are generally stand over the shrubs and forbs, not necessarily because they cover more area in a plant community. A fourth layer might be added where mosses are abundant in the ground cover, and some ultramafic soils have abundant ground cover of lichens (Fig. 8-4D).

Figure 4-3. TJ Howell Fen in the Klamath Mountains, a pitcher plant (*Darlingtonia californica*) fen with azalea bushes around the margin and a surrounding forest with an open canopy of Jeffrey pine trees.

Soils and plants alone are not completely functional geoecosystems. Arthropods, earth worms, nematodes, fungi, and microorganisms make essential contributions to the functioning of geoecosystems. The below ground biota are very important in the functioning of geoecosystems and deserve much more attention than they get. Mycorrhizal fungi, for example, are just as important for the maintenance of healthy plants in ultramafic soils as they are in nonultramafic soils (Southworth et al. 2013).

A broad range of ultramafic soils and plant communities from near the Arctic Circle to the Caribbean area is shown in Figure 1-4. The diversity is great.

C. Temporal Plant Community Trends

Like everything in the natural world, or universe, plant communities change over time. Although the current focus is on the communities that occupy ultramafic geoecosystems now, investigations of past plant distributions on ultramafic soils and changes expected in the future are acknowledged briefly here. Current climatic changes related to global warming may be more rapid than any comparable changes in the past.

4C1 Current vegetation and plant communities

Plant communities are continuously changing. Without major disturbances, local changes within geoecosystems may be ignored in describing the defining features and characteristics of the vegetation and plant communities in geoecosystems. Most of the serpentine plant community descriptions in the following chapters are qualitative, using personal judgement gained over many decades of field experience to assure that the plant community selections and descriptions are representative of reasonably mature communities rather than successional communities following major disturbances.

4C2 Substantive long-term changes

Daily and annual weather difference can be great; therefore climate is judged by average weather over one or more decades. It may have taken many decades, or centuries, of climate change to cause readily recognizable changes in mature plant communities. Through the Quaternary glacial stages, climates have varied substantially. Even in the past 12 or 15 ka, following the last major glaciation, there have been substantial plant community changes in some landscapes.

Figure 4-4. High Camp basin. A (left). A view across the basin, with Cement Bluff and Mt. Eddy on the right, with Mt. Shasta on the left. B (right). Cement Bluff and Bluff Lake.

The fossil record contains evidence of plants that lived in the past. Briles et al. (2011) examined pollen and noted charcoal in 15 ka cores from eight small lakes that are between about 1500 and 2290 m asl (above mean sea level) in the Klamath Mountains. Three of the lakes, including Bluff Lake (Fig. 4-4), are in basins with dominantly ultramafic rocks. Core samples older than 15 ka in some nonultramafic basins indicated occurrences of late glacial subalpine forest and sagebrush (*Artemisia* sp.). In nonultramafic basins the post-glacial vegetation fluctuated between dense forests with firs (*Abies* spp.) and white pines (*Pinus* spp.) and open forests with white pines, Cupressaceae, and huckleberry oak (*Q. vaccinifolia*). Hemlock (*Tsuga mertensiana*) has been present at the higher elevations and Douglas-fir (*Pseudotsuga menziesii*) has become more abundant recently. In dominantly ultramafic basins there has been much less variability in the structure (mainly plant density) of the forests; they have been dominated by yellow pine (*P. jeffrei*) and Cupressaceae. The Cupressaceae are mostly incense cedar (*Calocedrus decurrens*) now, with Port Orford cedar (*Chamaecyparis lawsoniana*) on the wetter soils. Plant species differences in nonultramafic basins indicate a vegetation shift 800 to 1000 m upward relative to current elevations following an abrupt warming after 11.7 ka, and cooling during the middle Holocene to shift species downward more than 200 m below current elevations.

Past changes in plant species distributions are clues to possible changes in the future, but the data can be incomplete. The investigation of Briles et al. (2011) revealed little about the forbs, or herbaceous plants. One-half a century after Whittaker (1960) recorded the plants in plots on granitic and ultramafic sites in the Klamath Mountains, Damschen and Harrison (Damschen et al. 2010) resampled the plants on those plots. They sampled herbaceous species, along with trees and shrubs, allowing them to make more complete predictions of vegetation trends, although not nearly as far into the future, or over such great climatic ranges, as the comparisons made with the data of Briles et al. (2011).

D. Serpentine Plant Species Adaptations

Plants that thrive on ultramafic soils tolerate low Ca availability (exchangeable Ca/Mg ratios <0.7 molar) and high Ni concentrations. These factors are inimical for most plants and may be largely responsible for the greater number of endemic plant species associated with ultramafic than with other soil parent materials. Although the substrate generally has minor influence on rock or soil mosses, a serpentine endemic moss species (*Pseudoleskella serpentinensis*) was reported from the Klamath Mountains (Malcolm et al. 2009).

Some plant species have intraspecies differences in adaptions to ultramafic soils (O'Dell and Rajakaruna 2011). Both serpentine tolerant and intolerant varieties of a species may occur in the same area where tolerant varieties occur on ultramafic soils and intolerant varieties occur off ultramafic soils (for example, yarrow, *Achillea millefolium*, Kruckeberg 1951); or the serpentine tolerant and intolerant varieties may be isolated from each other by occurring in different geographical locations (for example, bristly jewel flower, *Streptanthus glandulosus*, Kruckeberg 1957).

4D1 Low Ca/Mg ratios

Kazakou et al. (2008) attributed the low fertility of ultramafic soils primarily to low Ca/Mg ratios. Among the major plant nutrient elements, Ca is unique in that its functions are dependent on its concentrations in cell walls, rather than in the cytoplasm where it interferes with enzyme activities (Marschner 2002). Its main functions are in maintaining membrane stability and cell integrity. To prevent Ca interference in the cytoplasm, it is moved to vacuoles and the endoplasmic reticulum (Pilbeam and Morley 2007). Other cations can displace Ca from its active sites on the plasma membrane that encompasses cells, causing Ca-deficiency symptoms. The cations that replace Ca might be Na or K, but in ultramafic soils there is much more Mg, and divalent cations are more effective than monovalent cations in displacing other cations from sites where cations can be exchanged for one another. A mechanism that might enhance the ability of plants to tolerate ultramafic soils is the selective uptake of Ca, relative to Mg (Kazakou et al. 2008). O'Dell et al. (2006) found that the main difference between serpentine and nonserpentine varieties of some shrubs was in their different abilities to take up Ca relative to Mg and to move Ca from roots to shoots, rather than in differences in their abilities to tolerate Ni and other toxic "heavy" metals that are mainly first transition siderophile or lithophile elements.

Different plant species can respond quite differently to the limiting properties of serpentiune soils. In soils from the New Idria barrens of the California Coast Ranges, Lazarus et al. (2011) found that buckbrush (*Ceanothus cuneatus*) was more tolerant of the toxic transition elements and responded favorably to additional Ca and N, whereas yarrow (*Achillea millefolium*) and red brome (*Bromus madritensis* ssp. *rubens*) lacked favorable responses to added Ca and N in those soils with the toxic elements.

Foliage was sampled from yellow pine, Douglas-fir, and incense cedar trees on ultramafic and gabbro soils in mature, unmanaged forests (Alexander 2014b). Analyses of conifer foliage indicated that the yellow pine and incense cedar trees were able to maintain levels of foliar Ca in trees on ultramafic soils comparable to the foliar concentrations on gabbro soils, even though the ultramafic soils had much lower Ca contents and lower Ca/Mg ratios, but Douglas-fir trees on ultramafic soils had lower foliar Ca than those on gabbro soils. Even though the yellow pine trees on gabbro were ponderosa pine and those on ultramafic soils were Jeffrey pine, there were no significant differences in their foliar Ca and Mg concentrations.

4D2 Nickel accumulation

Most plants have Ni concentrations <1 μg/g. Serpentine plants generally have higher concentrations, up to 100 μg/g or more. Plants that contain Ni >1000 μg/g in their foliage are designated Ni-hyperaccumulators (Brooks et al. 1977, Baker and Brooks 1989, van der Ent et al. 2013). Only two plant species in western North America, mountain pennycress (*Noccaea fendleri*) in the Klamath Mountains and milkwort jewel flower (*Strepanthus polygaloides*) in the Sierra Nevada are certain to meet that criterion, and both are in the Brassicaceae. There are four varieties of milkwort jewelflower with different morphological characteristics. Of the two more common varieties, the one with purple sepals grows at higher elevations than the one with yellow sepals (Boyd et al. 2009). The majority of the more than 300 Ni-hyperaccumulators are in tropical areas, but there are numerous Ni-hyperaccumulators in the *Alyssum* and *Thlaspi* (or *Noccaea*) genera in the Mediterranean region.

Three western North American subspecies (abbreviation: subsp or ssp) of pennycress (*Noccaea fendleri*) are strict or facultative Ni-hyperaccumulators. *N. fendleri* ssp. *siskiyouense* and ssp. *californica* are Klamath Mountains endemics, and *N. fendleri* ssp. *glauca* is found on both ultramafic and nonultramafic soils over a much wider geographic range. A dodder (*Cuscuta californica*) parasitic on milkwort jewel flower was found to contain 800 μg Ni/g compared to only 11 μg Ni/g when growing on a nonhyperaccumulating plant in the same area (Boyd et al. 1999). Roberts (1992) reported hyperaccumulation of Ni in several Newfoundland plant species, but there has been no confirmation that they are hyperaccumulators, nor has a claim of hyperaccumulation in *Arenaria* (or *Minuartia*) *rubella* in western North America been confirmed

(Kruckeberg and Reeves 1995). There are many Ni hyperaccumulating plant species in Cuba. High concentrations of Ni may be responsible for excluding many plants from growth in ultramafic soils.

High concentrations of Ni in alpine pennycress (*Noccaea fendleri* ssp. *glauca*), which is a common Ni-hyperaccumulator in the Klamath Mountains, may deter insects from feeding on the plants (Boyd and Martens 1994, 1998). Jhee et al. (2005) restricted eight arthropods with different feeding modes to diets of milkwort jewelflower plants grown either on soils with high Ni concentrations, or on soils with low Ni concentrations. The high-Ni plants had mean leaf Ni of 6200 μg/g, and the low-Ni plants had mean leaf Ni of 38 μg/g. A root-feeding cabbage-worm avoided high-Ni plants and the plants were lethal for leaf-chewing insects such as grasshoppers, but the other arthropods fed on both high and low-Ni plants without any apparent ill effects. Those other arthropods included xylem and phloem feeders and two cell-disruptors, or homogenizers. Boyd (2009) proposed that insects with Ni concentrations >500 μg/g be considered high-Ni insects. There are at least 15 species of the high-Ni insects in South Africa, New Caledonia, and California, and the majority are bugs (Heteroptera, or Hemiptera) that feed on plants (Boyd 2009).

4D3 Serpentine endemism

The diversities of ultramafic soils and plants in western North America are great, especially from the Blue Mountains southward. Safford et al. (2005) have listed plant species that are more common on ultramafic substrates than on other substrates in California. Although ultramafic rocks occupy no more than 1% of the area in California, 12.7% of the endemic plant species are restricted to ultramafic substrates. More recently, Safford et al. (2020) revised the California endemic species inventory to 14.7% of them being serpentine endemics. Many species that have been identified as endemics are not limited strictly to ultramafic substrates but also some occur on gabbro and on other soils where they are not too much different from ultramafic soils.

The greatest amount of endemism in nontropical North America is in the Klamath Mountains (Harrison et al. 2006). Many endemic species might be expected there, because the Klamath Mountains have larger areas of ultramafic rocks than other regions in North America and the climatic diversity among the ultramafic areas there is very great. The ultramafic soils in the Klamath Mountains range from arid to humid (mostly with xeric SMRs) and from warm (thermic STR) to very cold (cryic STR). Soil properties closely related to endemism were found to be low exchangeable Ca/Mg from 5-15 cm soil depths, with molar ratios down to about 0.2 mol/mol (Harrison et al. 2006). With lower Ca/Mg ratios, the soils were assumed to be too harsh for many endemic plants to survive (Harrison et al. 2006).

Endemism in the Caribbean area, especially on the island of Cuba is even greater, much greater, than in the Klamath Mountains (Chapter 12). Borhidi (1991) estimated the total at 920 serpentine endemic plant species in Cuba. Isolation of the island on which about 7% (7500 km^2) of the area has ultramafic soils, much more than on other Caribbean islands, may be largely responsible for the large numbers of serpentine endemic species in Cuba. The Santa Elena Peninsula of Costa Rica with a relatively large area (about 340 km^2) of ultramafic soils has only one serpentine endemic plant species.

Note: Plants were identified taxonomically by reference to Baldwin et al. (2012), Hickman (1993), Hitchcock and Cronquist (1973), Hultén (1968), Rebman and Roberts (2012), Saint John (1963), and Wiggins (1980). Plant taxonomic names in current lists on the NRCS, US Department of Agriculture, site (plants.usda.gov) were assumed to be authoritative.

5 Ancient North America

Continental crust began to form about 4.0 Ga. Ultramafics in the ancient continental crust were volcanic rocks. Plate tectonics and the accretion of peridotite and serpentinite from oceanic crust was not initiated until Earth was more than a billion years old, possibly about 3 Ga..

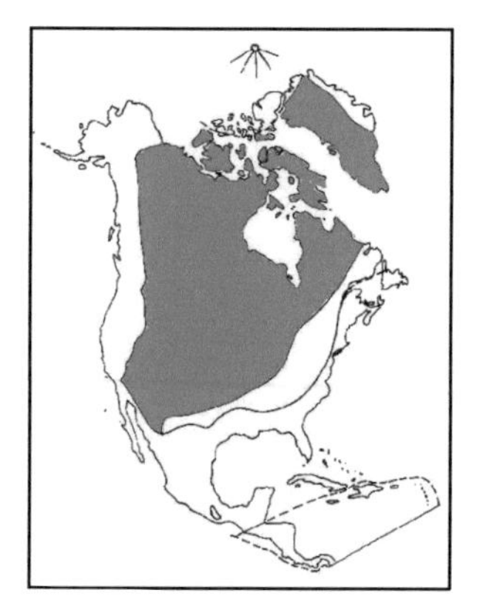

Ancient crust older than about 1.25 Ga is the core of North America. It consists of the Archean and early Proterozoic provinces that dominate the interior of North America west of the Appalachian Mountains and the Grenville province and east of the Paleozoic to recently accreted terranes that are along the margin of the Pacific Ocean on the west. It extends as far southwest as the controversial Mojave-Sonoran mega-shear (Fig. 2-1). These Archean and Proterozoic provinces are the ancient core that is commonly called the *craton* of the North American continent. The Archean provinces are called Slave, Rae, Hearne, Nain, Superior, and Wyoming. Proterozoic provinces preceding the late Proterozoic Grenville orogenies, are the Penokean, Mojave, Yavapai, and Mazatzal provinces (Condie 2005). The Proterozoic provinces are largely covered by Paleozoic and Cenozoic volcanic and sedimentary rocks and large areas of Phanerozoic plutonic rocks that have intruded them are exposed in the Rocky Mountains.

The Archean provinces consist primarily of greenstone belts developed from the metamorphism of basalt, picrite, and komatiite with intrusions of the more felsic TTG (tonalite-trondhjemite-granodiorite) rocks. There are no Archean ophiolites in North America, at least none like the familiar post-Archean forms. The main ultramafic rocks are the komatiites, which are volcanic rocks. The abundance of komatiites decreased as Earth cooled and they are sparse in the Proterozoic eon. There are no Phanerozoic komatiites other than on Gorgona Island, Colombia (Kerr et al. 1996), and possibly on the Nicoya Peninsula in Costa Rica (Alvarado et al. 1997). Other than komatiites, the only major ultramafic rocks of the Archean or early Preoterozoic provinces in North America are in the 2.7 Ga Stillwater complex in the Wyoming Province.

Following the Archean, the lack of komatiites and only minor continental additions of ultramafic rocks from plate tectonic activity through the early to middle Proterozoic left the Mojave, Penokean, Yavapi, and Matzatal Provinces nearly devoid of ultramafic rocks. Relatively significant, but small amounts of ultramafic rocks were intruded in the 1.2 Ga Muskox funnel-dike, or lapolith, that formed over the Mackenzie hotspot (Irvine and Smith 1967, Kerans 1983).

A. Komatiite

Ultramafic lava once flowed across the surface of Earth. Those volcanic flows were common billions of years ago, but ultramafic lava is no longer produced on Earth. The ultramafic volcanic flow rock is called *komatiite* after the Komati River in South Africa. On an oxide basis, which is the common practice in geology, komatiite contains about 45-50% silica (SiO_2) and 18-30% magnesia (MgO). Anhydrous komatiite melts at about 1600°C, which is somewhat higher than the melting temperature of basalt.

Komatiites were common in the Archean, became much less common through the Proterozoic as Earth cooled, and are absent through the Phanerozoic, except for 89 to 90 Ma komatiites on the Nicoya Peninsula and on Gorgona Island. Those Cretaceous komatiites were on the trailing edge of the Caribbean plate, which was moving eastward. North of the Caribbean plate, North American komatiites are found only in greenstone belts of the ancient craton. The most readily accessible of these greenstone belts is the Abitibi belt that stretches more than 300 km across Ontario into Quebec (Pyke et al. 1973, Jensen and Pyke 1982). Most of the rocks in the belt are covered with glacial drift. The typical North American locality of komatiite is Pyke Hill in the Abitibi belt (Arndt and Naldrett 1987). Pyke Hill is about 90 km east of Timmins Ontario; it has a very cold climate (Fig. 5-1)

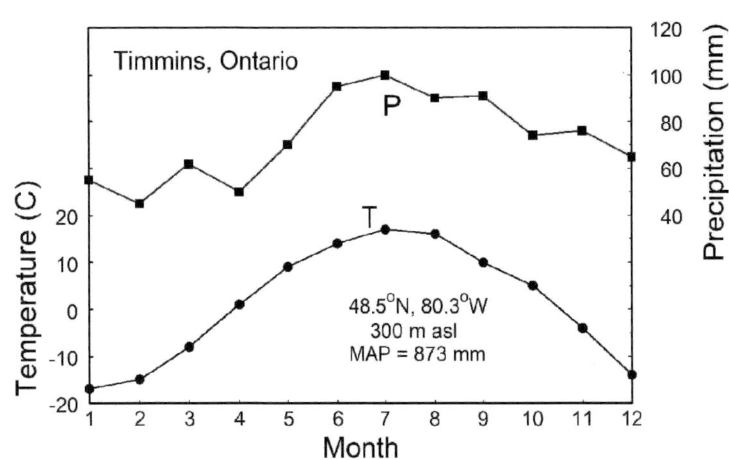

Figure 5-1. Mean monthly temperature and precipitation at Timmons in Ontario, Canada.

Pyke Hill is an exposure of komatiite that rises about 30 meters above an undulating plain of glacial drift (Fig. 5-2). Rock outcrop and shallow soils with shrubs and sparse to semi-dense tree cover are dominant. The most abundant vascular plants are Jack pine, white spruce, paper birch, dwarf huckleberry, sheep laurel, and mountain alder (Table 5.1). Moss (wavy dicranum and others) and lichen (*Cladina* spp.) cover is extensive on the rocks and soils, and moss is plentiful on trees, also. The surface of Pyke Hill is dominated by exposed bedrock and very shallow to shallow soils . A small swale a short distance below the summit of Pyke Hill contains a moderately deep soil that appears to be well drained, lacking redoximorphic features, even though willow is common on it and there are some sedges and rushes in the swale.

Figure 5-2. A view from the summit of Pyke Hill across the surrounding plain of glacial drift.

A very shallow soil was described on Pyke Hill (Fig. 5-3, Table 5.1). All stones, or pebbles, in the Pyke Hill soil appeared to be komatiite, but the silty soil texture indicates a major aeolian input, presumably from glacial outwash. Less than 5% of the fine-earth (soil particles <2 mm) is in the medium to very coarse sand fractions. Most of the fine-earth is finer: fine sand, more very fine sand, and much silt. Light minerals, mainly abundant rounded quartz and common weathered alkali feldpars, dominate the fine sand fractions of both E and Bs horizons; heavy nonmagnetic minerals (SG>2.89) are only about 10%, and magnetic grains are about 2% by weight. The heavy nonmagnetic grains are predominantly weathered pyroxenes with ragged edges, weathered olivine, and minor chlorite and tabular to rounded green hornblende. Compositions of the very fine sand and coarse silt fractions are similar, but there is less rounding of the mineral grains than in the fine sand fraction.

Although the R horizons of soils on Pyke Hill are ultramafic rock, the main soils have bleached E horizons and incipient Bs horizons (Table 5.1), unlike ultramafic soils. Also, the extremely acid Pyke Hill soils (Table 5.2) are much more acidic than ultramafic soils. The soil properties reflect the composition of aeolian accumulations of very fine sand and silt, rather than the underlying bedrock and coarse fragments in the soils. The cation exchange complex is dominated by aluminum. There is so little Ca and Mg that the low surface and subsoil Ca/Mg ratios of 0.5 and 0.3 mol/mol may not be the major factors in limiting the plants that can grow in these soils.

Figure 5-3 The Pyke Hill soil of Table 5.1. White graduations on the black tape are 10 cm long.

Table 5.1 A soil described on Pyke Hill, August 2010.

Location: 80.2°W, 48.6°N, 450 meters altitude (asl), Ontario, Canada
Classification/soil parent material: Lithic Cryorthent/komatiite and nonultramafic loess
Landform: gently sloping ridge subsidiary to the main ridge of Pyke Hill and slightly lower
Precipitation: 873 mm/year at Timmons, about 100 km to the west
Vegetative cover: trees (70%) - plentiful Jack pine (*P. banksiana*), common white spruce (*Picea glauca*) and paper birch (*Betula papyrifera*); shrubs (50%) - plentiful dwarf huckleberry (*Vaccinium cespitosum*), common sheep laurel (*Kalmia angustifolia*) and mountain alder (*Alnus viridis* ssp. *crispa*); sparse service berry (*Amelanchier* sp.), pin cherry (*Prunus pensylvanica*), and mountain-ash (*Sorbus* sp.); graminoids (trace) - mostly grasses; forbs (trace) - bunchberry (*Cornus canadensis*); sparse clubmoss (*Lycopodium clavatum* and *L. camplanatum*); mosses (5%) - wavy dicranum (*Dicranum undulatum*); lichens (20%) gray and star-tipped reindeer lichens (*Cladina rangiferina* and *C. stellaris*)
Surface rock and stoniness: slightly stony, extremely rocky

Oi 6 to 2 cm: shrub roots and pine needles

Oe 2 to 0 cm: partially decomposed leaves and roots

E 0 to 3 cm: very dark grayish brown (10YR 3/2, 5/2 dry) gravelly silt loam; stones and cobbles <1%, 20% pebbles; weak, very fine granular; soft, very friable, slightly sticky and slightly plastic; common very fine, fine, and medium roots; extremely acid (pH 4.0); abrupt, wavy boundary

Bs 3 to 16 cm: dark grayish brown (10YR 4/2, 6/2 dry) silt loam with few, fine, diffuse, reddish brown (5YR 4/5) mottles; stones and cobbles <1%, 5% pebbles; massive, structureless; friable, slightly sticky and slightly plastic; few very fine, fine, and medium roots; extremely acid (pH 4.2); abrupt, irregular boundary

R 16+ hard, coarsely jointed bedrock

Remarks: pH by bromocresol green indicator (glass electrode pH is lower, Table 5.2). A deeper (moderately deep) soil in a swale has a thicker bleached horizon and a thick strong brown (5-7.5YR 4/5 moist, 7.5 YR 6/6 dry) subsoil, and vegetative cover similar to that on the shallow soils, plus common willow (*Salix* sp.) and sparse Labrador tea (*Ledum groenlandicum*) and more graminoids other than grasses (sedges and rushes), but the soils do not appear to be poorly drained. The dominant mosses on the forest floor were identified by Dan Norris, Jepson Herbarium, but no mosses on the rocks and trees were identified.

Table 5.2 Texture by feel (field grade) and chemical analyses of the Pyke Hill soil.

Soil Horizon		Depth (cm)	Field grade	pH		Exchangeable cations			KCl extract. acidity	LOI 360° Celsius
				DW	KCl	Mg	Ca	Ca/Mg		
						$mmol_+/kg$		molar	mmol/kg	g/kg
-----	Oi/Oe	6-0	–	–	–	–	–	–	--	–
PkH1-1	E	0-3	GrSiL	3.4	3.1	2.5	1.2	0.5	41.3	47
PkH1-2	Bs	3-16	SiL	3.8	3.4	1.5	0.4	0.3	34.2	23

B. Stillwater Complex

Figure 5-4. Beartooth Plateau and Mountains southward from Iron Mountain on the Stillwater complex.

The Stillwater complex is in the Beartooth Mountains, just northeast of the Yellowstone volcanic plateau (Fig. 5-4). The complex is one of the world's largest layered mafic intrusions. Layers in the intrusion are differentiated by combinations of crystal fractionation and gravitational separation and by multiple injections of magma (McCallum 1996). The layers dip steeply to the north and strike west-northwesterly for about 48 km along the northern edge of the Beartooth Plateau. The topmost layers have been lost by erosion, leaving a vertical thickness of about 6 kilometers in the complex. The complex is divided into (1) a basal chill zone (mostly fine-grained gabbro) followed upward by layers of gabbro, norite, and feldspar-pyroxenites, for a total thickness about 0.7 km, (2) a layered ultramafic zone (alternating layers of dunite, peridotite, and bronzite-pyroxenite) approximately 1 km thick, and (3) a layered mafic zone, approximately 4 km thick, with alternating layers of norite, gabbro, troctolite, and anorthosite. A striking feature of the complex is regular and persistent layering that is remarkably similar to sedimentary bedding. The complex has been considered to be the product of fractional crystallization of basaltic magma trapped in a chamber with a horizontal floor. The Stillwater complex contains large reserves of chromite and associated platinum and palladium. There are large areas of barren rock, boulder or block fields, and talus on the Beartooth Plateau and in ravines cut into the Plateau. The climate is very cold, even at the Mouat Mine, more than 1000 m lower than the Beartooth Plateau.

Near the Ben Bow Mine (about 2500 m asl), the soils observed over bronzite (an orthopyroxene with more Fe than most enstatite and less than most hypersthene) are Alfisols (Haplocryalfs) and Inceptisols (Cryepts) dominated by young stands of lodgepole pine (*P. contorta*) with sparse subalpine fir (*Abies lasiocarpa*) and white pine (*Pinus* sp.) and with grouse whortleberry (*Vaccinium scoparium*) and serviceberry (*Amelanchier* sp.) in the understory. Low juniper (*J. communis*) and kinnikinnick (*Arctostaphylos uva-ursi*) are present on the more shallow soils.

Figure 5-5. A subalpine forest on the north side of Iron Mountain. The main plant species are Engelmann spruce (*Picea engelmannii*), whitebark pine (*Pinus albicaulis*), subalpine fir (*Abies lasiocarpa*), and grouse whortleberry (*Vaccinium scoparium*).

The main plants about the summit of Iron Mountain (3075 m asl) are mountain-avens (*Dryas octapetala*), cespitose phlox (*Phlox pulvinata*), bistort (*Polygonum bistortoides*), mountain lupine (*Lupinus lyallii*), cinquefoil (*Potentilla* sp.), and sedge (*Carex* sp.). Although quartzite is exposed at the very summit of Iron Mountain, the same plants that are on quartzite near the summit occur over pyroxenite around the summit. Most of the same plants dominate alpine slopes below the summit, except that mountain-avens is missing and a cespitose sandwort (*Minuartia obtusiloba*) is present. Whitebark pine trees are scattered sparsely across the alpine to subalpine slopes. On the north side of Iron Mountain, forests with spruce (*Picea engelmannii*), whitebark pine (*Pinus albicaulis*), and subalpine fir (*Abies lasiocarpa*) in the overstory and grouse whortleberry (*Vaccinium scoparium*) in the understory prevail (Fig.5-5). A very cold, well-drained ultramafic (pyroxenite) Alfisol (Mollic Haplocryalf) at 3060 m asl described in a meadow on Chrome Mountain had a cover of about 35% forbs, 20% tufted hairgrass (*Deschampsia cespitosa*), 20% sedges (*Carex* spp.), and scattered whitebark pine trees. The dominant forbs were miniature lupine, bistort, and sandwort.

6 Proterozoic Grenville Orogen

The oldest orogenic episodes in which appreciable ultramafic rocks were accreted to ancestral North America were a series of late Proterozoic orogenies that all together are called the Proterozoic Grenville orogen (Tollo et al. 2004). These orogenies were on the margins of a Proterozoic supercontinent called Rodinia. Grenville age accreted terranes have been scattered worldwide following the breakup of Rodinia near the end of the Proterozoic eon. Proterozoic rocks of the Grenville Province in North America are present from Labrador through Quebec and Ontario, with outliers in the Adirondack Mountains and inliers in the Appalachian Mountains (McLelland et al. 2010). Rocks of Grenville age are exposed in central and western Texas (Mosher 1998), northern Mexico (Ruiz et al. 1988), and the Oaxaca complex of southern Mexico (Fig. 2-1). Strata of the Oaxaca complex were involved in the Zapotecan orogeny (Sedlock et al. 1993), which was contemporaneous with the Grenville 0rogeny. Grenville age rocks of the Oaxaca complex were not accreted to the North American craton until the Paleozoic Era (Ortega-Gutierrez et al. 1995).

A. Grenville Province

The Grenville Province extends from Labrador to Lake Huron. This area was covered by the Laurentide ice sheet. Preglacial soils have been stripped from much of the area and Pleistocene glacial drift covers large areas, minimizing the exposures of ultramafic soils. Anorthosite, an ancient rock that Phanerzoic terranes lack, is extensive in the Marcy Massif of the Adirondack Mountains, but there are no notable exposures of ultramafic rocks in those mountains. Highlands on our moon are composed of anorthosite (density, SG<2.8) where it floats over the ultramafic rocks (Carlson 2019).

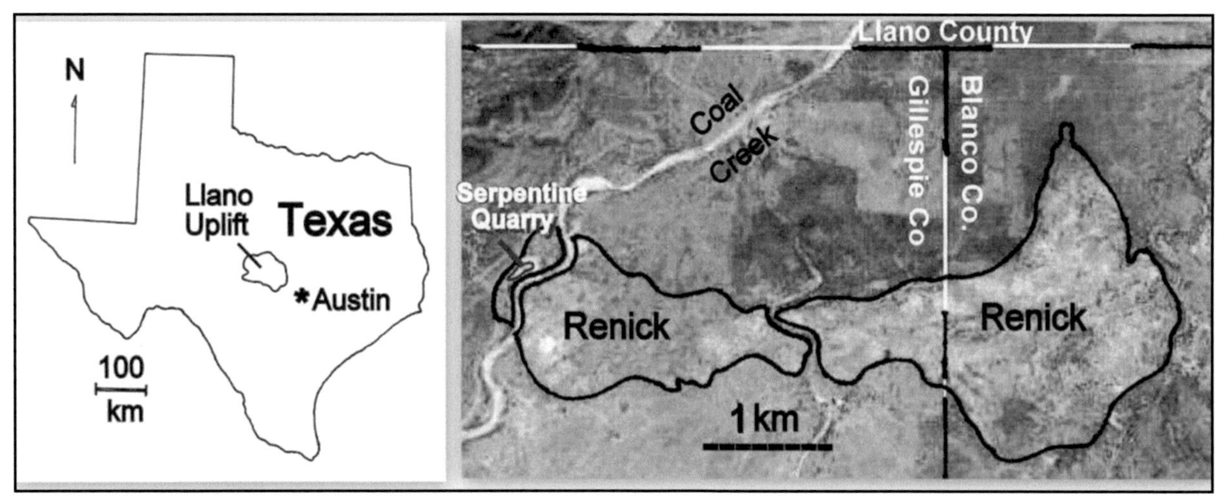

Figure 6-1. The Grenville age terrane in the southern part of the Llano Uplift. Renick soil polygons were mapped in county soil surveys of the Natural Resources Conservation Service.

B. Central Texas

The ultramafic rocks in central Texas are in the Llano uplift (Fig. 6-1). They are serpentinized peridotite, or serpentinite, that was accreted to Laurentia about 1.2 to 1.3 Ga. The ultramafic body is about 6 km long and 0.5 to 2 km wide, as reported by Garrison (1981). The serpentinite is predominantly lizardite, with accessory magnetite; relict peridotite minerals are sparse. The latitude is about 30.5°N. The summers are hot and the soils are seasonally dry (Fig. 6-2).

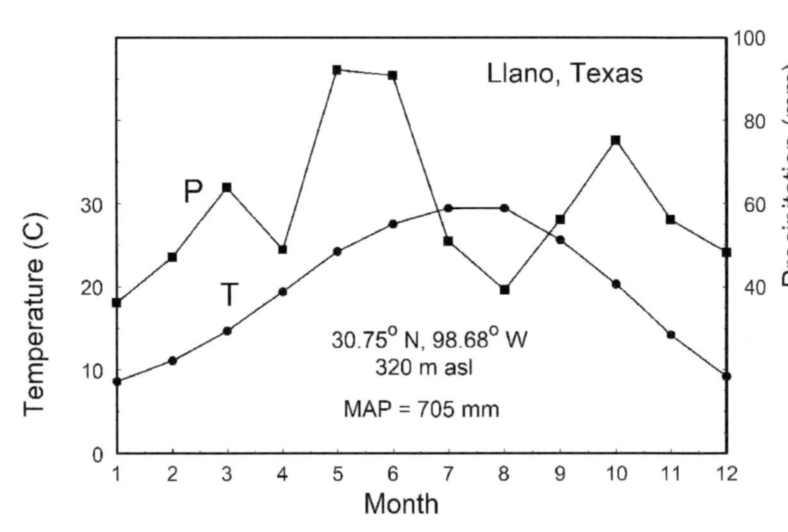

Figure 6-2. Mean monthly temperatures and precipitation at Llanos, about 40 km NNW of the serpentinized peridotite.

Ultramafic soils of the Llano Uplift in central Texas are in Gillespie (316 ha) and Blanco (309 ha) counties (Allison et al. 1975, Dittemore and Allison 1979). This is a hilly area (Fig. 6-3) with altitudes about 300 to 420 m asl. The ultramafic soils are mostly gently sloping, but with steep slopes along Coal Creek (Fig. 6-4). They are stony soils in a map unit of the Renick series of Mollisols (clayey,

Figure 6-3. A view from a serpentinite quarry across Coal Creek to the landscape of the Renick soil.

smectitic, thermic Ruptic-Lithic Haplustolls, Table 6.1), with inclusions of rock outcrop, lithic argillic Mollisols (Lithic Argiustolls), and moderately deep argillic Mollisols (Pachic Argiustolls). Soils on steeper slopes are similar to the Renick soils, but in clayey-skeletal families (Fig.6-4). A student of BL Allen (Maoui 1966) found that the fine clay in the Renick soils is mainly smectite and the coarse clay is predominantly serpentine and chlorite, with the serpentine more concentrated in the C horizon and the chlorite in the upper horizons. Some quartz, amphibole of the tremolite-actinolite series, and talc were found in the Renick soils and may be from schists that are closely associated with the serpentinite, although tremolite, or actinolite, and talc are commonly formed by alteration of the peridotite from which the serpentine is produced by serpentinization. Bangira (2010) found 7 g/kg Cr in the surface to 5 g/kg in the subsoil of a shallow soil in Gilespie County, and all was trivalent (CrIII). He attributed the lack of hexavalent Cr (CrVI) to the presence of 26 g/kg organic C (about 5% organic matter) in the surface to 19 g/kg in the subsoil (organic matter is about 50% organic carbon, C). The dominant serpentine vegetation was oak-juniper-mesquite savanna.

Figure 6-4. A steep slope along Coal Creek and cacti on that slope.

Nixon and McMillan (1964) studied the plant communities at six sites in Texas, including two in northeastern Gillespie County—one on ultramafic soils and the other on granitic soils. An oak-mesquite savanna on the granitic soils was represented by post oak (*Quercus stellata*), cedar elm (*Ulmus crassifolia*), and honey mesquite (*Prosopis glandulosa*, var. *glandulosa*), with little bluestem (*Schizachyrium scoparium*), switchgrass (*Panicum virgatum*), Indiangrass (*Sorghastrum nutans*), and sideoats grama (*Bouteloua curtipendula*). The woody vegetation was scanty on the ultramafic soils, but the four grasses mentioned for the granitic soils were common, along with hairy grama (*Bouteloua hirsuta*) and curly-mesquite (*Hilaria belangeri*), on the ultramafic soils, although Indiangrass was sparse. Other common species on the ultramafic soils that were mentioned by Maoui (1966) were Virginia liveoak (*Quercus virginiana*), catclaw (*greggii*), Ashe's juniper (*Juniperus ashei*), Texas persimmon (*Diospyros texana*), algerita (*Mahonia trifoliolata*), prickly pear (*Opuntia* sp.), composite forbs in the *Thelesperma* and *Chaetopappa* genera, and several grass species. Among the grass species were Texas wintergrass (*Nassella leucotricha*), fall witchgrass (*Digitaria cognata*), Texas grama (*Bouteloua rigidiseta*), purple threeawn (*Aristida purpurea*), sand dropseed (*Sporobolus cryptandrus*), plains lovegrass (*Eragrostis intermedia*), silver bluestem (*Bothriochloa saccharoides*), and Hall's panicgrass (*Panicum hallii*). Also, there was some wild buckwheat (*Eriogonum* sp.) and croton (*Croton* sp.). Buckley's yucca (*Yucca constricta*) is another species that is common on the steep ultramafic soils along Coal Creek (Fig. 6-4).

Table 6.1 Renick soil series, excerpts from the official soil series description (NRCS 2009)*.

Hor.	Depth	Munsell Color		Structure[1]	Roots	Acidity (pH)	Boun -dary[2]
	cm	dry	moist	grad; size; form	abund, size		
A1	0-20	10YR 4/2	10YR 3/2	m; f; sbk	many, fine	neutral	cs
A2	20-38	10YR 4/3	10YR 3/3	m; f,m; sbk	common, fine and medium	neutral	cw
A3	38-48	10YR 4/3	10YR 3/3	m; f,m; sbk	few fine	neutral	cw
R	48-50+	5Y 6/2 to 5Y 7/2		hard, fractured serpentinite; few roots in fractures			

*The complete description is available on the USDA, Natural Resources Conservation Service website.
[1] Structure: grade - m, medium; size - f, fine, and m, medium; form - sbk, subangular blocky.
[2] Boundary: cs, clear and smooth; cw, clear and wavy.
Soil consistence (all horizons): hard when dry, friable when moist, and sticky when wet.
Rock fragments: few, hard in A1 and A2; many (about 20%), soft in A3 horizon.

7 Eastern North America

Appalachian orogenic area
Newfoundland to Alabama

The ultramafic rocks of eastern North America are in mountains from Newfound through the Appalachians and on the Piedmont that is between the mountains and the Atlantic coastal plain. The area from Newfoundland to Staten Island has been glaciated and is designated the northern Appalachian area. The unglaciated area south from Staten Island is designated the southern Appalachian. Climates are very cold to cool in the northern Appalachian and warm to cool in the southern Appalachian area. Snow falls in all of the areas, but it is a continuous winter ground cover only in the northern Appalachian and the higher mountains in the southern Appalachian area.

Soil and plant community differences in the northern Appalachian are commonly related to soil drainage differences. Soils in the southern Appalachian areas are more commonly well drained, they are more weathered, and they commonly have subsoil accumulations of clay. There are great differences in plant cover and plant species from the cooler north to the warmer south.

Although the evolution of terranes that occur in the Appalachian area began in the Neoproterozoic era, only Paleozoic and later events are decipherable with any detail. Near the end of the Neoproterozoic era, when the supercontinent Rodinia broke up, one of the fragments was Laurentia, the precursor of the North American continent. Laurentia was separated from Gondwana and Baltica by the Iapetus Ocean. Early in the Paleozoic, the Iapetus Ocean began to close. Baltica impinged upon Laurentia early in the Paleozoic causing the Caledonian orogeny from eastern Greenland (Hendriksen 2008, Higgins et al. 2008) to northern Europe. The early to late Paleozoic Taconian, Acadian, and Alleghanian orogenies, from Newfoundland to Alabama accompanied the closing of the Iapetus Ocean. Coincident with the southern Appalachian Alleghanian orogeny, the South American part of Gondwana impinged upon Laurentia in the Ouachitan orogeny. No more than minor amounts of ultramafic rocks were exposed in the Ouachitan orogeny of southeastern North America, or in the Innuitian (or Ellesmerian) orogeny of Arctic North America, and no ultramafic rocks have been added to eastern North America since the Paleozoic era.

A. Geology

Early in the Paleozoic era, mantle-derived rocks from ocean spreading centers and magmatic arcs were accreted to Laurentia, along with pieces of Laurentia that had rifted away from it upon the breakup of Rodinia. This event has been called the Taconian orogeny and the autochthonous and allochthonous accreted terranes of it are designated Laurentia-Iapetus (Fig. 7-1). During the Taconian and subsequent Paleozoic orogenies, rocks inland from the accreted terranes were displaced and deformed and they are designated Laurentia (Fig. 7-1). Ganderia, Carolinia, and Avalonia rocks that were accreted during the early to middle Paleozoic Salinian and Acadian orogenies (Eyles and Miall 2007) resemble African components of Gondwana. Meguma, the last piece added to Laurentia, resembles parts of north Africa. Late in the Paleozoic era, Africa closed upon the southern Appalachian area in the Alleghanian orogeny and South America closed upon southeastern North America in the Ouachitan orogeny. This aggregation of continents formed Pangea, the last supercontinent, about 0.25 Ga. Pangea broke up early in the Mesozoic and no exotic terranes have been accreted to eastern North America following the Paleozoic era.

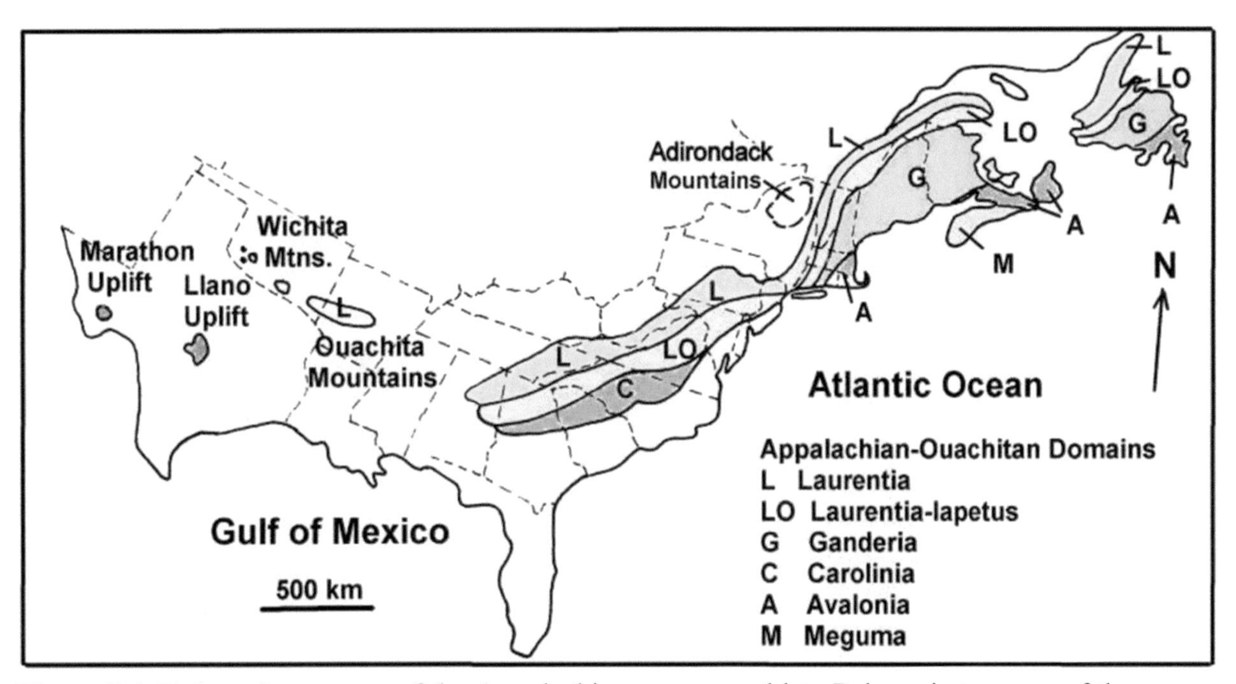

Figure 7-1. Paleozoic terranes of the Appalachian orogen and late Paleozoic terranes of the Ouachitan orogen. Units of the Appalachian orogen are in accord with those of Hibbard et al. (2006). Terranes of the Llano and Marathon Uplifts are Grenville age. Thr C polygon (Carolinia) includes all of the Piedmont, both ultramafic and nonultramafic strata.

7A1 Northern Appalachian ultramafics

Appalachian ultramafic rocks are most abundant in the Laurentia-Iapetus domain, which in the northern Appalachian area has been called the Dunnage domain. These rocks are most concentrated along faults on the western boundary of the Laurentia-Iapetus area (Fig. 7-1), such as along the Baie Verte-Brompton and Taconic lines (Hatch 1982, Williams and St-Julien 1982) from Newfoundland through southeastern Quebec, Vermont, western Massachusetts and Connecticut (Fig. 7-2, Larrabee 1971). Ophiolites with ultramafic components are readily recognized in Newfoundland and Quebec, but deformation and metamorphism make ophiolites more difficult to identify south of the Thetford Mines, Asbestos, and Mt. Orford ophiolites in Quebec. The general sizes of ultramafic bodies diminish southward along the Baie Verte-Brompton and Taconic Lines, and the rocks in western New England are characteristically almost completely serpentinized with some talc schist and talc-magnesite or magnesite aureoles.

Ultramafic rocks are less abundant in Ganderia and Avalonia than in the Laurentia-Iapetus domain. There are none in Meguma. There are small bodies of ultramafic rocks along the Connecticut Valley-Gaspé trough and on the Bronson anticlinorum in central New England and adjacent Canada (Lyons et al. 1982), on Deer and Little Deer Isles in Penobscot Bay (Wing 1951), and on Boil Mountain and in the Jim Pond formation in western Maine (Boudette 1982).

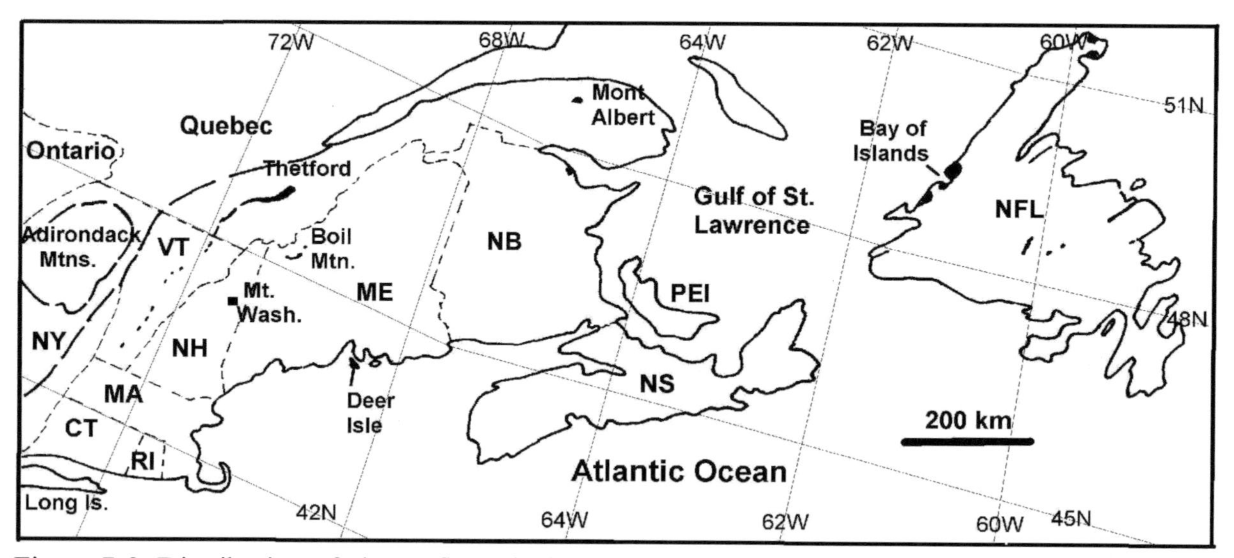

Figure 7-2. Distribution of ultramafic rocks in the northern Appalachian area. They are mainly in Newfoundland (NFL), on Mont Albert, and from Theford south through Vermont (VT).

7A2 Southern Appalachian ultramafics

The largest southern Appalachian exposures of ultramafic rocks are in the Baltimore mafic complex. Ultramafic rocks of the Baltimore mafic complex are in a basal unit of serpentinized dunite containing chromitite and in an overlying gabbroic unit approximately 5 km thick that contains many cumulus peridotite layers (Hanan and Sinha 1989). Rocks of the Baltimore mafic complex are mainly in three large blocks from southeastern Pennsylvania to near Baltimore in Maryland. Also, lenses and pods of the ultramafic rocks in the Soldiers Delight belt occur in the Wissahickon formation that is in or adjacent to the Baltimore complex.

South of the relatively large Baltimore mafic complex, ultramafic rocks are most concentrated along the Brevard fault zone on the eastern margin of the Blue Ridge Mountains and in associated foothills and highlands (Fig. 7-3). In western North Carolina and northern Georgia, there are many small and lenticular bodies of dunite conforming to the regional rock foliation (Misra and Keller 1978). Serpentinization is evident around the margins of these bodies, and soapstone is common. Alteration of dunite and harzburgite in the lenticular bodies increases toward the northeast, and ultramafic rocks and associates in the Abemarle-Nelson belt of Virginia are mainly amphibole-chlorite schist, serpentinite, soapstone, and other rocks derived from the alteration of peridotite and dunite. The Abemarle-Nelson is a belt less than I km wide with several narrow bands less than 2 km to more than 20 km long in the foothills of the Blue Ridge Mountains, parallel to the inner (inland) margin of the Piedmont.

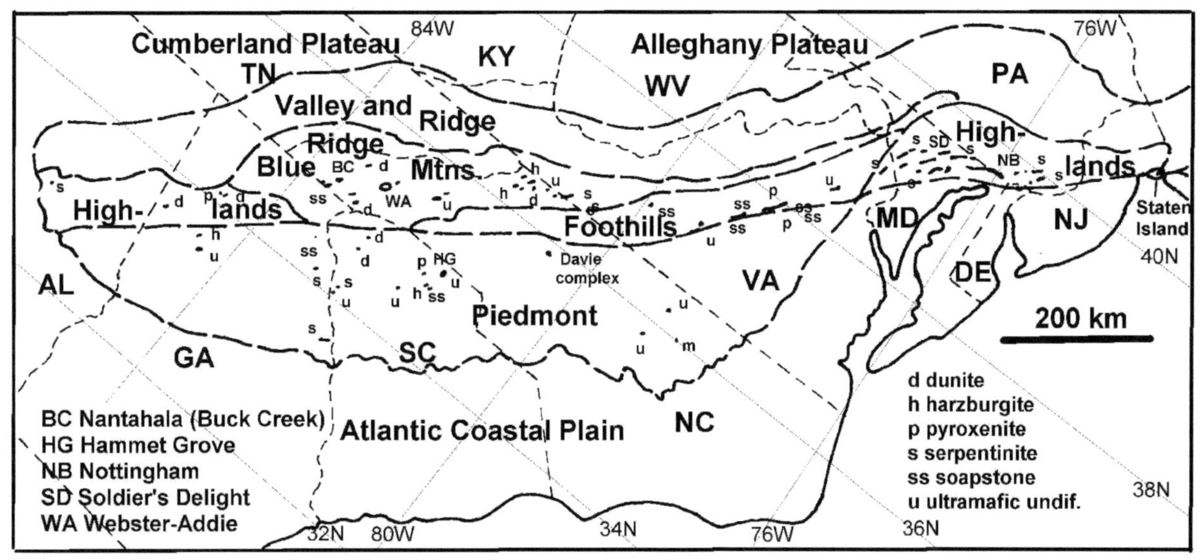

Figure 7-3. Distributions of ultramafic rocks and physiographic units in the southern Applachian area. Rock locations and petrographic designations are after Larabee (1966).

Small lenses of altered Laurentia-Iapetus domain ultramafic rocks occur throughout the inner Piedmont (Misra and Keller 1978), which is separated from Carolinia on the outer Piedmont by the central Piedmont shear zone (Hibbard et al. 2006). Several elongated ultramafic bodies up to 2 km wide are scattered across the outer Piedmont, where they are less common than on the Inner Piedmont.

The relatively large Hammett Grove suite along the edge of the King Mountain belt, near the central Piedmont shear zone, may represent an ophiolite in which the ultramafic rocks have been metamorphically altered to soapstone and serpentinite (Mittwede 1989). It is exposed for about 11 km along the strike, with a common width of 200 to 300 m and a maximum width of 1 km (Butler 1989). Iredell soils (Hapludalfs with "mixed" mineralogy) were mapped on metapyroxenite and metagabbro of the Hammet Grove suite (Jones 1962, Mittwede 1989). Ultramafic rocks of the Charlotte and King Mountain belts along the central Piedmont shear zone are mainly minor constituents of mafic intrusive complexes that are generally located near faults (Butler 1989).

B. Physiography

Mountains of the Appalachian region are generally highest near the western, or northwestern, margin of the Applachian orogen area and decrease in elevation toward the Atlantic coastal plain. The White Mountains in New England (around Mt. Washington, Fig. 7-2) are exceptions; they are the highest in the northern Appalachians. The southeastern margin of the Appalachian orogenic area is obscured by the Atlantic coastal plain. Toward the north, most of the coastal plain is under water, below sea level.

7B1 Northern Appalachian physiography

On the northwest margin of the Appalachian orogen, the Notre Dame Mountains of Gaspésie and the Long Range Mountains in Newfoundland rise abruptly above the St. Lawrence lowland and the Gulf of St. Lawrence. From that margin, the topographic highs are represented by one or more planes that slope southeast to the Atlantic coastal plain, which is offshore of Newfoundland and Nova Scotia. Some geologists have suggested that the summits representing these planes are remnants of ancient peneplains.

South from the Notre Dame Mountains, the western margin of the northern Appalachians is bound by low valleys (valleys in unit L of Fig. 7-1). These are the Champlain Valley west of the Green Mountains and the Hudson Valley west of the Taconic Mountains. The highest summits of New England are in the White Mountains of New Hampshire (Mt. Washington, 1889 m, Fig. 7-2) and in Maine (Mt. Katahdin, 1580 m), rather than in the Green and Taconic Mountains (Denny 1982). A massive delta west of the Taconic Mountains, the Catskill delta, was produced from

Appalachian sediments after Acadian uplift in the middle Paleozoic era.

All of the northern Appalachians, south to Long Island, were covered by ice during the last major glaciation, although there may have been some unglaciated nunataks on Newfoundland and the Gaspé Peninsula. For sometime after the Laurentide continental glacier had melted, icecaps persisted on Newfoundland and the Gaspé Peninsula. Ice cover slowly disappeared from the Gaspé Peninsula between 10 and 13.5 ka. Permafrost is still present above about 1200 m on Mont Jacques-Cartier (1270 m), but the active layer above permafrost is about 6 m deep (Gray and Brown 1979).

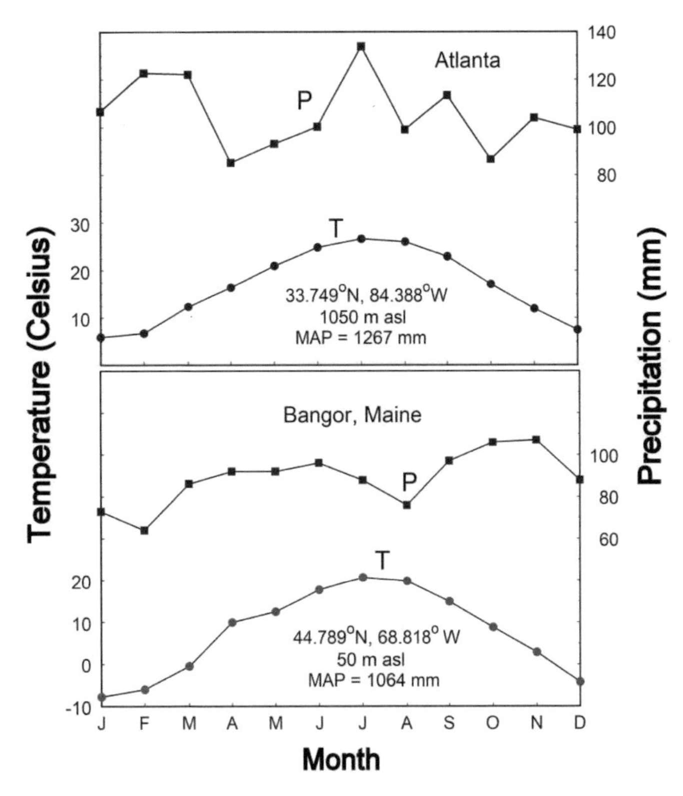

Figure 7-4. Mean monthly temperatures and precipitation at Bangor ME and Atlanta GA.

7B2 Southern Appalachian physiography

The highest parts of the southern Appalachians are in the Blue Ridge Mountains (Fig. 7-3) in North Carolina (Mitchell Mountain, 2037 m) and highs diminish both southward and northward from Mitchell Mountain (Hack 1982). The highest mountains in northern Virginia, Maryland, and Pennsylvania are in the Valley and Ridge province, rather than in the Blue Ridge Mountains. As in the northern Appalachians, the topographic highs are represented by one or more planes that slope southeast to the Atlantic coastal plain. In the southern Appalachian, however, there are two abrupt drops in the northwest–southeast topographic trend. One is at the transition from the Blue Ridge Mountains and associated foothills and highlands near the Brevard zone to the much lower, undulating to hilly Piedmont, and the other is at the Fall Line from the Piedmont to the Post-Paleozoic sedimentary cover on the Atlantic coastal plain.

C. Climate and Hydrology

Climates of the Appalachian Mountains are continental, because prevailing west winds minimize the ameliorating effects of the Atlantic Ocean. The mean annual temperatures range from near zero to about 16°C. The frost-free season is about 3 or 4 months in the north to 7 months in the south. Mean annual precipitation is mostly in the 750 to 1500 mm range (Fig. 7-4), but up to about 2000 mm at the higher elevations in the southern Appalachian Mountains. Summer temperatures are cool in Newfoundland to hot at the lower elevations in the southern Appalachian area. The winter precipitation is mostly snow in the northern Appalachian and at high elevations in the southern Appalachian area; it is rain or snow at lower elevations in the southern Appalachian area. Drought is expected in the Appalachian area only on shallow soils.

Stream runoff is fairly uniform from month to month, with peaks when most of the winter snow melts. The peaks are small, about March, in the southern Appalachian area and large peaks, about April, in the northern Appalachian area. Stream flow continues through the summers; only small headwater streams become dry during summers.

Cleaves et al. (1974) monitored the water chemistry of a small stream in the Soldiers Delight ultramafic area of Maryland. They sampled four times a year for a year. The mass of dissolved constituents in stream discharge exceeded that of particulate losses from the small ultramafic watershed. Watershed losses of constituents dissolved in stream water were dominated by bicarbonate (HCO_3^-), magnesium (Mg^{2+}), and silica (presumably silicic acid, H_4SiO_4). Cleaves et al. (1974) compared the stream losses to those from a small felsic schist watershed. The total concentration of dissolved constituents was much lower in the nonultramafic watershed and the cations were present in different proportions (Table 7.1). Other than Si^{4+}, Mg^{2+} was the dominant cation in water from the ultramafic watershed, whereas Na^+ was the predominant cation in water from the felsic schist watershed.

Table 7.1 Mean concentrations of anions, cations, and silicon (reported as silica) in base flow from physiographically similar ultramafic and nonultramafic (felsic schist) watersheds at Soldiers Delight (Cleaves et al. 1974).

Bedrock	Anions ($mmol_-/L$)				Cations ($mmol_+/L$)				Silica
	HCO_3^{1-}	NO_3^{1-}	Cl^{1-}	$SO4^{2-}$	Ca^{2+}	Mg^{2+}	Na^{1+}	K^{1+}	SiO_2
ultramafics	2.02	0.02	0.09	0.47	0.15	2.49	0.10	0.005	20.0
felsic schist	0.13	–	0.06	0.03	0.07	0.06	0.15	0.02	0.3

Units of silica content are mg/L.

In contrast to the usual bicarbonate waters of nonalkaline and moderately alkaline streams, travertine ($CaCO_3$) is deposited from very strongly alkaline spring water and seeps in some ultramafic areas of Newfoundland (Roberts 1992). The genesis of travertine generating spring water is discussed in Chapter 2.

D. Soils and Vegetation

Ultramafic soils of the Appalachian area range from very cold to warm. From north to south they have cryic (very cold), frigid (cold), mesic (cool), and thermic (warm) soil temperature regimes (Soil Survey Staff 1999). Ultramafic soils of the Appalachian region have udic (moist) soil moisture regimes, with aquic conditions common in Newfoundland and on the Gaspé Peninsula. Weathering, leaching of basic cations and silica, and the oxidation of iron are major soil-forming processes in the freely drained ultramafic soils. Smectites are produced where silica and magnesia accumulate and, in soils of the southern Appalachian area; interlayered chlorite/vermiculite minerals are common. In well-drained soils of the northern Appalachian area, podzolization is a major soil-forming process, but only in nonultramafic soils, or in soils with parent materials that are not entirely ultramafic materials. In both ultramafic and nonultramafic soils of the southern Appalachian area, incorporation of organic matter as humus into surface A horizons and accumulation of clay in argillic or kandic B horizons are major soil-forming processes. Because Spodosols (or Podzols, FAO/ISRIC/ISSS 1998) do not develop in purely ultramafic materials, or in ultramafic materials that lack a cover of nonultramafic materials, the more developed ultramafic soils in the northern Appalachian area are Inceptisols (Cambisols and Umbrisols). Alfisols and Ultisols (Luvisols, Lixisols, and Acrisols) prevail as the more developed soils of the southern Appalachian area. (Table 7.2). Another factor is Pleistocene glaciation that covered the northern Appalachian south to Long Island and northern New Jersey. Soils in the southern Appalachian are beyond the limit of Pleistocene continental glaciation, consequently many of them are much older than any in the northern Appalachian area.

Table 7.2 The most-developed and common well-drained soils in the Appalachian area.

Soil Temperature Regime	Nonultramafic Soils		Ultramafic Soils	
	Soil Order	Diagnostic Hor.	Soil Order	Diagnostic Hor.
Glaciated Terrains				
Cryic (very cold)	Spodosol	spodic	Entisol or Inceptisol	none or cambic
Frigid (cold)	Spodosol	spodic	Inceptisol	cambic
Mesic (cool)	Inceptisol	cambic	Inceptisol	cambic
Nonglaciated Areas				
Mesic (cool)	Ultisol	argillic	Alfisol	argillic
Thernic (warm)	Ultisol	argillic, kandic	Alfisol	kandic

Reed (1986) has botanical descriptions from many of the major ultramafic localities in eastern North America, including small pipes and dikes of ultramafic rocks from the mantle of Earth that are present west of the Appalachian Mountains from Canada to Arkansas. Ultramafic areas at the higher latitudes and those at higher elevations in the northern Appalachian area lack trees. Southern Appalachian ultramafic areas are commonly in pine and oak savannas that are perpetuated by burning. Eastern North America lacks confirmed Ni hyperaccumlulating plants (Pollard 2016), although some have been reported for Newfoundland.

The vegetation emphasis in this section is on major features and functioning of plant communities and geoecosystems. Greater details on taxonomic and evolutionary botanical aspects, lichens, and bryophytes can be found in Rajakaruna et al. (2009). Briscoe et al. (2009) and Medeiros et al. (2014) found large diversities of bryophytes on ultramafic strata of the Deer Isles, Maine.

Pope et al. (2010) listed several forbs and ferns that are restricted to ultramafic rocks and soils in eastern North America. Apparently, *Cerastium arvense* ssp. *velutinium* var. *villosum* (Caryophyllaceae) is the only plant that is recognized as a serpentine endemic in eastern North America.

7D1 Northern Appalachian glaciated area

Ultramafic rocks are more concentrated in western Newfoundland than in any other Appalachian area (Fig. 7-2). These ultramafic rocks are mostly in ophiolites, along with gabbro and basalt, that occupy about 3% (0.3 Mha) of the land in Newfoundland (Roberts 1992, Fig. 7-5). The ultramafic soils and plants on the Mont Albert plateau of the Gaspé Peninsula are similar to those in Newfoundland (Sirois and Grandtner 1992). Most of the soils have sandy loam and loam textures and are affected by cryoturbation. Sorted polygons and stone stripes are common.

The two main ultramafic soils of Newfoundland, which are mostly in glacial drift and colluvium, are the Roundhill series of well-drained Orthic Regosols (Soil Classification Working Group 1998) on slopes and the Blomidon series of somewhat poorly drained Gleyed Regosols on bottom land (Table 7.3; Roberts 1980, 1992). According to the USDA's Soil Taxonomy, the soils are Entisols and Aquolls, poorly drained Mollisols (Soil Survey Staff 1999). The soil reactions are neutral (near pH 7), although some organic layers are more acid (pH <6.5). Roberts (1992) reported that some of the Caryophyllaceae (*Arenaria humifusa* and *Minuartia marcescens* and marginally *Lychnis alpina*) and Asteraceae (*Solidago hispida* and *Packera paupercula*) were nickel hyperaccumulators. He found that Ni availability increased as the soil pH decreased.

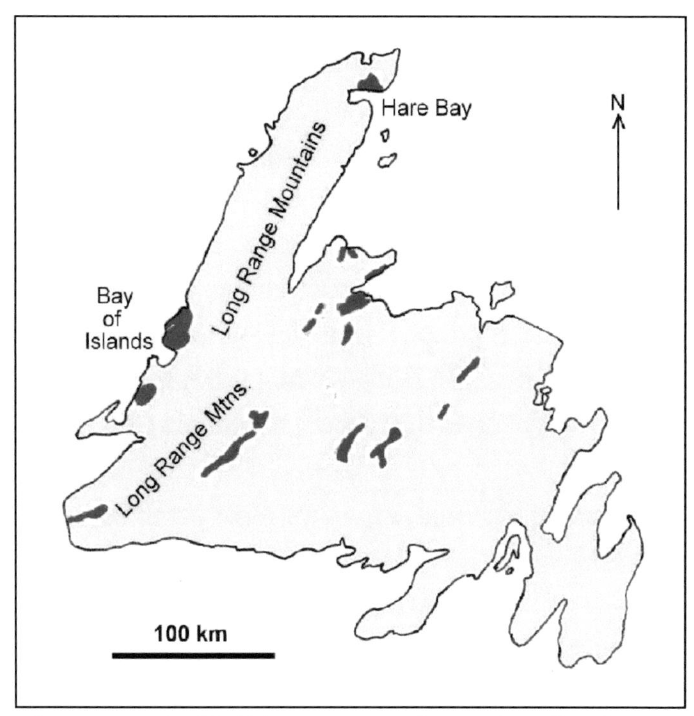

Figure 7-5. Ophiolitic terranes of Newfoundland.

Near Table Mountain in the Long Range Mountains of western Newfoundland, Dearden (1979) described several plant communities from scarps and from talus at the bases of scarps, across patterned ground to fens and meadows. He noted the marked contrast from fir-spruce forests common on Newfoundland to the relatively barren ultramafic landscapes. A fern (*Adiantum aleuticum*) prevailed in rock crevices at the bases of the scarps. Some of the scarps that are below the 712-m summit of Table Mountain have foot-of-scarp soils with relatively high Ca/Mg ratios and alkaline reaction (pH 8) that might be traced to the seepage of water from serpentinized peridotite, although Dearden did not mention this linkage. The alkaline soils supported small tamarack (*Larix laricina*) trees and low and creeping juniper (*Juniperus communis* and *J. horizontalis*) shrubs. Stone stripes and sorted polygon borders were colonized first by *Racomitrium lanuginosum* (hoary rock moss), followed by Lapland rosebay (*Rhododendron lapponicum*), several species of willow (*Salix* spp.), and herbs in the pink family (Caryophyllaceae). The nonstony polygon centers were colonized by chamaephytes such as *Silene acaulis*, *Armeria maritima* ssp. *sibirica*, *Diapensia lapponica*, and *Saxifraga oppositifolia*. The wetter polygons were colonized by cespitose bulrush (*Scirpus* sp.), star sedge (*Carex echinata*), and highland rush (*Juncus trifidus*). The mosses were big redstem (*Pleurozium schreberi*), feather moss (*Hylocomiun splendens*), and Bonjean's dicranum moss (*D. bonjeanii*). Some small depressions had bog rosemary (*Andromeda polifolia* var. *glaucophylla*), sweetgale (*Myrica gale*), and creeping juniper shrubs, also. Isolated hummocks were colonized by scrub alder (*Alnus viridis* ssp. *crispa*), followed by other dwarf tree species. Banks at the edges of solifluction terraces were colonized by prostrate tree species such as bog birch (*Betula pumila*), and arctic willow (*Salix arctica*), and shrubs, including alpine blueberry (*Vaccinium uliginosum*), cinquefoil (*Dasiphora fruticosa*), and common juniper. Organic soils in the fens were generally very thick (depth >2 m) and acid (pH <6). They were dominated by sedges (*Carex* spp.), with red fescue (*F. rubra*), cottongrass (*Eriophorum* spp.), and rushes (*Juncus* spp.). The wet meadows were also dominated by many species of Cyperaceae.

Two plant species, *Cypripedium parviflorum* and *Dryas integrifolia*, were found on Newfoundland only in areas with travertine (Bouchard et al. 1978). This travertine is the same as that reported by Roberts (1992).

Both Dearden (1979) and Roberts (1992) reported the dominance of kaolinite in clay fractions of drained soils and smectite in poorly drained soils, but they did not show data to verify the presence of kaolinite, which is unlikely to occur, instead of serpentine, in the ultramafic soils. De Kimpe and Zizka (1973) found that serpentinized dunite containing serpentine, brucite, and magnetite weathered by way of interstratified minerals to produce chlorite and smectite in an Orthic Melanic Brunisol (Eutrocryept) at Asbestos, Québec. De Kimpe et al. (1987) found that all primary minerals in pyroxenites at Mont Saint-Bruno, Province of Quebec, weathered to produce smectites.

Sirois and Grandtner (1992) related the ultramafic soils and serpentine plant communities on the Mont Albert plateau (Fig. 7-6A) to slope characteristics and drainage. As in Newfoundland, the initial pioneer community was *Racomitrium lanuginosum*. Dwarf birch (*Betula glandulosa*), alpine blueberry, and Lapland rosebay were characteristic shrubs. Apparently *Cetraria laevigata* (striped Iceland lichen) replaces the moss as the shrub layer develops. Soils on well-drained sites developed from Cryic Regosols to Orthic Humic Regosols and Eutric Brunisols. Well-drained Mollisols on amphibolite of the plateau were Dystric Brunisols. The krumholz soils were Orthic or Gleyed Ferro-Humic Podzols (not Spodsols) with black spruce (*Picea mariana*) on ultramafic soils and with balsam fir (*Abies balsamea*) on sites less exposed to wind and on amphibolite. The Podzols (Canadian classification) on ultramafic strata are not Spodosols, because they lack albic E horizons. Soils on poorly drained sites were Orthic Humic Gleysols, commonly with Labrador tea (*Ledum groenlandicum*) plant communities containing much sedge, grass, and bulrush, or with club-moss (*Selaginella selaginoides*). Very poorly drained soils were Rego Gleysols and organic soils with bulrush–moss plant communities. The bulrush–moss communities were dominated by cespitose bulrush (*Scirpus* sp.) and either campylium moss (*C. stellatum*) or *Sphagnum* spp. The former contains Labrador tea and bog rosemary, and the latter bulrush–moss

Figure 7-6. Ultramafic landscapes of the northern Appalachian area. A. Mont Albert. B. Pine Hill summit. C. Pine Hill sideslope.

community contains small cranberry (*Vaccinium oxycoccus*), *Drosera rotundifolia*, and *Maianthemum trifolium*. Organic soils, which are not extensive on the Mont Albert plateau, were predominantly Terric Humisols (Terric Cryosaprists). Some of the Histosols were on small mounds, or hummocks, among Histic Cryaquolls. There may have been permafrost beneath the mounds (BA Roberts, 2008, personal communication).

Table 7.3 Subsoils properties of ultramafic soils in Newfoundland and in the southern Appalachian area, including organic carbon to 0.5 and one meter depths.

Soil Order or Suborder	State Prov.	Lat. °N	Sub. Hor.	Color (moist)	Clay (g/kg)	Ca/Mg (molar)	pH	OC (kg/m^2)	
								0.5 m	1 meter
Orthent	NFL	49.2	C	2.5Y 5/4	195	0.09	7.0	2.7	–
Aquoll	NFL	49,2	Cg	2.5Y 3/2	256	0.05	7.0	11.0	–
Hapludalf	MD	39.4	Bt	10YR 5/4	–	0.03	–	5.2	6.3
Hapludalf	NC	35.1	Bt	7.5YR 5/8	343	0.03	6.1	4.3	5.8
Hapludalf	NC	35.1	Bt	5YR 4/6	553	0.09	5.9	6.4	9.8
Kanhapludalf	NC	35.4	Bt	10R 3/6	526	0.02	5.4	7.0	9.8

The Orthent and Aquoll are the Roundhill and Blomidon soils and the Kanhapludalf is in the Ellijay series. All of the soils have udic (or aquic in Aquolls) soil moisture regimes.

Ultramafic rocks do not appear to be very favorable substrates for saxicolous (rock) lichens, but some are found on ultramafic rocks or nearby amphibolite exposures on the Gaspé Peninsula (Sirois et al. 1988). Asbestos mine wastes in the Thetford Mines area are unfavorable substrates for all vascular plants. Even after adding fertilizer and organic amendments, Moore and Zimmerman (1977) found critically low N and Ca concentrations in plants grown on the wastes.

Much of the ultramafic terrain in New England is covered by glacial drift and lacks distinctively serpentine vegetation. Zika and Dann (1985) listed some plants that are characteristic of ultramafic habitats from Québec through New England to Pennsylvania. They are *Asplenium trichomanes* (maidenhair spleenwort), *Campanula rotundifolia*, *Cerastium arvense*, *Deschampsia cespitosa* (tufted hairgrass), and *D. flexuosa* (wavy hairgrass). Largeleaf sandwort (*Moehringia macrophylla*) and *Adiantum aleuticum* are ubiquitous on rocky dunite outcrops in Vermont.

Rajakaruna et al. (2009) have summarized the current knowledge of serpentine plant distributions in the northern Appalachian area, and Harris et al. (2007) and Briscoe et al. (2009) have compiled lichen and bryophyte species lists for the Pine Hill exposures of serpentinized peridotite on Little Deer Isle, Maine. Characteristic soils of the Pine Hill body, which is a clinopyroxene-rich peridotite that is about 60% altered to serpentine, are very shallow Entisols and shallow Inceptisols. The plant cover ranges from bare rock and grass (mainly *Danthonia spicata*, *Deschampsia flexuosa*, and *Festuca filiformis* on the summit of Pine Hill, Fig. 7-6B) to a white spruce-white cedar/dwarf juniper-bayberry plant community on side slopes that lack glacial drift (Fig. 7-6C). An ultramafic body of dunite and harzburgite on Deer Isle has a serpentinite margin where blocks were quarried until it was determined that rock fracturing was too great for a commercial operation.

7D2 Northern Piedmont Highlands

On Staten Island, Parisio (1981) described and sampled a soil on serpentinite in a hilly driftless area between Wisconsin moraines and Cretaceous sedimentary rocks of the Atlantic coastal plain. It had a brown (7.5YR 4/4), very stony loam cambic horizon with only 8% clay and an exchangeable Ca/Mg ratio of 0.15 mol/mol. The pH was 6.2 in distilled water and 5.7 in KCl solution. Serpentine and smectite were the main clay minerals, with some talc, mica, and goethite. Quartz and feldspar were found in very fine sand from the solum, but not in the parent rock. The soil was a mesic (STR) Inceptisol (Eutrudept). The plant community was a little bluestem (*Schizachyrium scoparium*) prairie, with predominantly little bluestem, roundleaf greenbrier (*Smilax rotundifolia*), and blackberry (*Rubus allegheniensis*).

The serpentine vegetation of the Nottingham barrens in southwestern Pennsylvania and the Soldiers Delight barrens in Maryland (Fig. 7-7), both on ultramafic rocks of the Baltimore complex, are the most thoroughly studied serpentine areas of the southern Appalachian area. The term "barrens" has been used for treeless prairies, or savannas, in otherwise forested parts of eastern North America (Tyndall 1994). They have relatively good plant cover, compared to the barrens of western North America.

Figure 7-7. Virginia pine (*P. virginiana*) savanna at Soldier's Delight, with blackjack and post oaks (*Q. marilandica* and (*Q. stellata*) and grasses, largely little bluestem, threeawn, prairie dropseed, and Indiangrass.

Ultramafic soils of the Baltimore complex have been mapped in the moderately deep Chrome (Typic Hapludalfs), very deep Aldino (Typic Fragiudalfs), and deep Conowingo (Aquic Hapludalfs), and Calvert (Typic Fragiaqualfs) series. About 1700 ha of the Chrome and 750 ha of the Conowingo soil have been mapped in Chester and Delaware counties, Pennsylvania, and about 2750 ha of the Chrome and 245 ha of the Conowingo soils were mapped in Baltimore, Cecil, Harford, Howard, and Montgomery counties, Maryland. The area of very deep Aldino soils is greater than that of the Chrome and Conowingo soils, but the Aldino soils were mapped on gabbro as well as on ultramafic strata, and much loess added to the Aldino and Calvert soils has given them properties atypical of ultramafic soils (Rabenhorst and Foss 1981).

Rabenhorst et al. (1982) studied four pedons of Alfisols (Lithic, Typic, and Aquic Haploxeralfs) on ultramafic strata of the Baltimore complex. Some nonultramafic loess had been incorporated into the soils, mainly in the surface soils where molar exchangeable Ca/Mg ratios were 0.8 or more in three of the soils, but ratios < 0.1 mol/mol prevailed in the subsoils (Table 7.3). Soil textures were silt loam in the surfaces and silt loam or silty clay loam in subsoils. Whereas pH values generally decrease with depth in nonultramafic soils of the northern Piedmont, the pH values in these ultramafic soils increased from very strongly acid in A horizons to pH 6.6 to 6.8 in subsoils. Serpentine minerals were present in the coarse clay and silt fractions, but absent from the fine clay. Smectites dominated the clay fractions in one soil and chlorite and vermiculite, were dominant in the other soils. Most of the vermiculite was interstratified with chlorite.

Soils on serpentinite of a small Soldiers Delight watershed monitored by Cleaves et al. (1974) are moderately deep to shallow and lack the saprolite that is common over bedrock of igneous and metamorphic rocks in this part of the northern Piedmont. Most of the silicon and magnesium from the weathering of ultramafic materials is leached from the soils, with relatively small amounts of these elements forming smectites and with minor amounts of silica precipitated in the weathered bedrock to form silica boxwork. Chemical denudation exceeded particulate losses from the serpentinite watershed (Cleaves et al. 1974).

Harris et al. (1984) found talc in saprolite weathered from a metamorphosed ultramafic rock in the Piedmont. The talc resisted weathering sufficiently to appear in the solum of the soil, which was a very deep Ultisol (Hapludult, or Acrisol), above the saprolite.

A century or more ago, the ultramafic areas of the northern Piedmont were mostly prairies with trees only in ravines and riparian areas where they were not burned frequently. The dominant grasses were little bluestem, threeawn (*Aristida purpurascens*), prairie dropseed (*Sporobolus heterolepis*), and Indiangrass (*Sorghastrum nutans*); the main trees were blackjack oak (*Quercus marilandica).* and post oak (*Q. stellata*); roundleaf greenbrier (*Smilax rotundifolia*) was a common shrub or vine; and common serpentinophile forbs were chickweed (*Cerastium arvense*) and serpentine aster (*Symphyotrichum depauperatum*). Following the cessation of burning and grazing, the prairies have been invaded by Virginia pine (*P. virginiana*) and Virginia juniper (*J. virginiana*) (Fig. 7-7), plus pitch pine (*Pinus rigida*) northeast of the Susquehanna River. Some of the ultramafic areas are currently managed for the preservation of scarce serpentinocoles (Tyndall 1992) such as *Agalinis acuta*, *Phemeranthus teretifolius*, and *Linum sulcatum*. According to

Reed (1986), *Selaginella rupestris* (spike-moss), *Phemeranthus teretifolius*, and *Phlox sublata* are good serpentine indicator plants on the Baltimore complex, but he has reported collecting them on nonultramafic soils farther south, also.

The main effects of pine trees on the Chrome soils are increased O-horizon thickness, lower surface soil pH, and decreased exchangeable Mg, resulting in increased Ca/Mg ratios (Barton and Wallenstein 1997). Increased soil depth was reported, but the pine trees may have become established in deeper pockets of soil, or in less stony soils where a depth measuring rod would be more likely to measure depth to bedrock rather than to a stone above bedrock. Also, increased plant detritus under the trees could cause an apparent increase in soil depth if O-horizons are included in the soil depth. The availability of K and several other plant nutrient elements increase in at least the organic surface horizons following pine invasion, making the soils more favorable for the invasion of other plant species on the ultramafic soils (Hochman 2001).

Burgess et al. (2015) have affirmed the succession from xeric oak-savanna to the encroachment of Virginia pine and juniper and ultimately to mesophytic forest. They found that the succession is more advanced on deeper (shallow to moderately deep) soils than on more shallow soils near rock (serpentinized peridotite) outcrops.

It was once commonly thought that the availability of soil water was the main factor limiting plant diversity and growth on the ultramafic soils. Hull and Wood (1984) measured summer soil water and oak tree xylem potentials. They concluded that the availability of water does not appear to be the factor allowing blackjack oak and post oak to replace white oak (*Quercus alba*) and black oak (*Q. velutina*) on ultramafic soils. The consensus has shifted toward the Ca/Mg ratio as being a major factor controlling which plants will grow on the ultramafic soils.

Terlizza and Karlander (1979) collected algae from ultramafic soils at Soldiers Delight. The algae were found to be Cyanophyta (some of which fix nitrogen), Chlorophyta, and Chrysophyta. At this level of classification, the soil algae composition is similar to that of nonultramafic soils.

Panaccione et al. (2001) found lower ectomycorrhizal fungi diversity on serpentine plots on ultramafic soils at Soldiers Delight than on nearby nonultramafic soils. They collected *Cenococcum geophilum* Fr. isolates from Virginia pine seedlings in both ultramafic and nonultramafic soils and found that the *C. geophilum* isolates from ultramafic sites were genetically more similar to each other than to isolates from local or distant nonultramafic sites.

A beetle (*Diabrotica cristata* Harris) that is seldom found along the Atlantic coast, but is common farther west, was found to be abundant on the Goat Hill and Nottingham barrens and present at Soldiers Delight (Wheeler 1988). The main host plant for the larvae, which are called rootworms, is big bluestem (*Andropogon gerardii*) in the Midwest and assumed to be little bluestem on serpentine prairies and savannas of the Baltimore complex.

7D3 Blue Ridge Mountains and the Southern Piedmont

Small areas of residual ultramafic soils 15 m deep, or deeper, derived mainly from peridotite and dunite, are present in the Blue Ridge Mountains (Worthington 1964). Very deep, strongly weathered soils are present on the Piedmont as far north as Staten Island, also, but they are not as deeply weathered as those in the Blue Ridge Mountains. Weathering and leaching removed most of the Mg and much of the Si from the old ultramafic parent materials and Cr, Fe, Co, and Ni have been concentrated 5- to 7-fold in the soils (Worthington 1964).

Analyses of soapstone from Abemarle and Fairfax counties, Virginia, and very deep soils in it, show that most of the Mg and much of the Si have been lost by weathering and leaching, leaving soils that are about one-third Al- and Fe-oxides (Merrill 1906). This soapstone is metamorphosed peridotite, or pyroxenite, and is composed of talc and chlorite, with inclusions of tremolite.

Only one ultramafic soil series has been established by the NRCS on the small exposures of ultramafics in the Blue Ridge Mountains. It is a very deep Alfisol in the Ellijay series of fine, ferruginous, kandic Alfisols (Rhodic Kanhapudalfs, or Acrisols). About 250 ha of the Ellijay soils have been mapped on peridotite, or dunite, in Jackson County, North Carolina. Major features of Rhodic Kanhapludalfs are a kandic horizon and red colors (hues 2.5YR, or redder, and color values moist of 3 or less and dry of 4 or less). A kandic horizon is a horizon of clay concentration, like an argillic horizon, except that the kandic horizon has low cation-exchange capacity (CEC <160 mmol+/kg of clay at pH 7). At the type locality for the Ellijay series, there is 9.8 kg of organic C in the upper meter of soil (NRCS pedon S85NC-099-004; Table 7.3).

Radford (1948) described seven plant communities on dunite and serpentinized peridotite at nine sites in North Carolina and one site in Georgia. He did not describe the soils. A pine-bluestem community was the most distinctively serpentine one; the others were similar to plant communities on other kinds of rocks. In the pine-bluestem community, Virginia pine and pitch pine were the dominant trees, with subdominant post oak and southern red oak (*Q. falcata*). Jersey tea (*Ceanothus americanus*), deerberry (*Vaccinium stamineum*), azalea (*Rhododendron calendulaceum*), and mountain laurel (*Kalmia latifolia*) were common shrubs, and big and little bluestem and panic grass (*Panicum* spp.) were the most common grasses.

Ogg and Smith (1993) studied two deep soils from the Blue Ridge Mountains in North Carolina derived from altered peridotites containing tremolite, or actinolite, and chlorite, with no olivine, and a moderately deep soil from the Piedmont in South Carolina derived from a hornblende-bearing pyroxenite. The soils were Alfisols (Hapludalfs). The main clay minerals were found to be interstratified chlorite/vermiculite, kaolinite, and talc in the North Carolina soils and smectite and kaolinite in the South Carolina soil; perhaps the hornblende added enough extra Al for the formation of the kaolinite. Goethite was the main iron-oxide mineral.

Ultramafic soils with vegetation that is most distinctively different from that on other kinds of soils in the Blue Ridge Mountains are not deep(?) soils, according to Mansberg and Wentworth (1984). They described a pitch pine plant community on the Buck Creek (or Nantahala) ultramafics in Clay County, North Carolina. Sparse, but ubiquitous, saplings and understory trees were hemlock (*Tsuga canadensis*), white oak, red maple (*Acer rubrum*), and sassafras (*Sassafras albidum*), and the shrubs were serviceberry (*Amelanchier arborea*), deerberry, ninebark (*Physocarpus opulifolius*), and cat greenbrier (*Smilax glauca*). The most common forb was prairie groundsel (*Senicio plattensis*, or *Packera plattensis*); and big and little bluestem were ubiquitous grasses. Prairie groundsel and prairie dropseed (*Sporobolus heterolepis*) were noted because of their disjunct distributions from midcontinental prairies to Appalachian ultramafic soils.

No specifically ultramafic soils were mapped on the Nantahala (Buck Creek) ultramafics in Clay County (Thomas 1998). Subsequently, three soils presumed to be among the more dominant ultramafic soils in the Nantahala area have been described and sampled (Table 7.4). All three soils had enough subsoil (B horizon) clay for argillic horizons, and two of them were argillic Mollisols, but there were no clay coatings in the soil on talc-chlorite schist and it was classified as an Inceptisol (Fig. 7-8A). Fine sand from the C horizon of the Inceptisol contained flaky talc-like grains and traces of tremolite. X-ray diffractograms revealed chlorite and talc in the C horizon, but only chlorite in the B horizon. Some of the large amounts of Mg extracted from the Inceptisol (Table 7.4) may have been removed from talc and chlorite upon leaching with NH_4-acetate to remove the exchangeable cations. The Nantahala soils on peridotite (Fig.1-4D and 7-8B) have accumulated enough organic matter for mollic epipedons, although the mollic epipedon is below an extremely acid A1 horizon in the moderately deep Mollisol (Fig. 7-8B). Well-drained Mollisols (including Lithic and Typic Argiudolls) are sparse in the southern Appalachian, and may occur only with certain parent materials that are not geographically extensive (Stanley Buol, North Carolina State University, Raleigh, 2008, personal communication).

Figure 7-8. Ultramafic soils in Nantahala (Buck Creek) landscapes.
A. A shallow Inceptisol in a cleared and burned savanna.
B. A moderately deep Mollisol in a pitch pine-white oak forest.

Table 7.4 Nantahala ultramafic soils, Nantahala National Forest, Blue Ridge Mountains.

Hor.	Depth (cm)	Color (moist)	Texture (feel)	LOI (g/kg)	Exch. Cations (mmol+/kg)			Ca/Mg (molar)	pH
					Ca	Mg	Acidity*		
Lithic Eutrudept in a cleared and burned pitch pine–red maple savanna (Fig. 7-8A).									
Oi	1-0	loose leaves							
A	0-11	10YR 3/2	L	110	21	232	37	0.09	6.0
Bw	11-23	7.5YR 4/4	CL	—	9	655	18	0.01	6.4
C	23-38	N 8/0	L	—	8	346	8	0.02	6.7
R	38+	hard, slightly fractured talc-chlorite schist							
Typic Argiudoll in a pitch pine–white oak forest (Fig. 7-8B)									
Oi/Oe	5-0	loose leaves/fragmented, partially decomposed, slightly matted leaves							
A1	0-9	10YR 3/2	L	267	17	149	211	0.11	4.3
A2	9-22	10YR 3/3	GrL	55	8	73	28	0.11	5.8
A3	22-36	10YR 3/3	GrCL	79	9	124	35	0.08	6.0
Bt	36-70	7.5YR 4/6	GrCL	—	7	71	25	0.10	6.2
Cr	70-98	6YR 5/8	soft, massive weathered bedrock						
R	98+	hard, slightly fractured peridotite							
Lithic Argiudoll in mixed oak–red maple–hemlock forest (Fig. 4-1D)									
Oi/Oe	3-0	loose leaves/partially decomposed, slightly matted leaves							
A1	0-2	10YR 2/2	L	—	—	—	—	—	—
A2	2-9	10YR 3/2	L	154	39	166	79	0.24	5.3
A3	9-22	10YR 3/3	L	79	17	97	48	0.17	5.8
Bt1	22-29	10YR 3/2	GrCL	—	19	124	46	0.15	6.0
Bt2	29-34	7.5YR 5/6	vGrCL	—	20	75	19	0.26	6.2
Cr	34-36	5YR 4/6	soft, weathered bedrock, pH 6.9						
R	36+	hard, massive dunite							

* Exchange acidity at pH 7.0.

Milton and Purdy (1988) sampled the foliage from several species of trees growing on ultramafic soils in the Buck Creek and the Webster-Addie districts in the Blue Ridge Mountains . White oak leaves accumulated the most nickel, about 400 to 700 μg/g from five sites at Buck Creek, but they had Ni <200 μg/g from sites at the Webster-Addie district.

Soils of the Iredell series, which are Mollisols that developed in mafic materials, rather than from ultramafic parent material, are known to support many plant species that are not found on adjacent soils of the Piedmont. Locally, areas of Iredell soils have been called blackjack soils because they have larger proportions of blackjack oak and post oak, relative to white oak, than surrounding soils. Dayton (1966) found that the characteristic plants on the Iredell soils were those common on calcareous soils, in xeric habitats, or with prairie affinities. He did not find any distinctive soil physical or chemical properties, such as the low exchangeable Ca/Mg ratios of ultramafic soils, that would cause the Iredell soils with mafic parent materials to have highly distinctive vegetation.

Seven soils in Georgia, North Carolina, South Carolina, and Virginia identified as Iredell when sampled were analyzed in a National Resources Conservation Service laboratory. These soils are generally moderately deep to saprolite, but numerous roots penetrate deeply into soft saprolite. The soil parent materials were not identified in the pedon descriptions, but one of them was called "multicolored gneiss," and minerals in the fine sand fraction were found to be mostly hornblende and feldspar with some quartz. Six of the seven soils had exchangeable Ca/Mg ratios <1.0 in their subsoils, and four of those had ratios <0.7. The soil reaction ranges widely from pedon to pedon, but can be generalized as slightly acid in the surface, neutral in the subsoil, and slightly alkaline in the saprolite. The Iredell soils with mafic parent material are intermediate between those with more silicic and those with ultramafic parent materials, and apparently the vegetation has characteristics that might be expected on soils with either silicic or ultramafic parent material.

E. Summary

Ultramafic soils and soil-vegetation relationships

Broad ranges of ultramafic soils and vegetation occur from Newfoundland to Alabama. The ultramafic soils are distinguished more by their mineralogy and chemistry than by their morphology. The main morphological differences from nonultramafic soils are that the cold ultramafic soils lack the distinctive albic (bleached) horizons that are characteristic of Spodosols (Podzols), and the greater amounts of secondary iron oxides in warm, strongly weathered ultramafic soils commonly give them redder colors than nonultramafic soils. Also, in some areas where the ultramafic soils are commonly Mollisols, the nonultramafic soils have no mollic epipedons. Although botanists commonly concentrate on the shallow ultramafic soils that have the most distinctive serpentine vegetation, ultramafic soils range from shallow to very deep. There is no evidence that ultramafic soils are in general more shallow, or have lower available-water capacities (AWCs), than nonultramafic soils over large areas with many different kinds of soils. In a comparison of available-water capacities (AWCs) from all soils in the Klamath and Shasta-Trinity National Forests, California, no differences between ultramafic and nonultramafic soils were evident (Alexander et al. 2007a). The organic carbon in the upper meter of ultramafic soils in eastern North America (Table 7.3) is comparable to that in other soils. The main chemical features that distinguish ultramafic soils are very high Mg contents and the first transition elements that are more strongly siderophiles than chalcophiles, especially relatively high Cr, Co, and Ni, compared to other soils. It is very low exchangeable Ca/Mg ratios (Table 7.3), and high Ni contents, that impose the greatest limitations to plants on ultramafic soils. Any effects of relatively low K and P in ultramafic soils are masked by the effects of very high Mg and low Ca concentrations (Alexander et al. 2007a). Nitrogen deficiency symptoms may appear upon fertilization of ultramafic soils with other elements (Moore and Zimmerman 1977), but plants respond favorably to N additions on most soils, regardless of parent materials. Ultramafic soils are not particularly low in N, compared to many other soils (Alexander et al. 2007a).

Although the lesser density and low stature of serpentine vegetation are the most readily visible differences from nonserpentine vegetation, the plant species distributions are generally different, also. Many plants on ultramafic soils with low available-water capacities occur mostly, or only, on ultramafic soils, and some are isolated from their common geographic ranges (disjunct distributions). These unique vegetation distributions make ultramafic soils very interesting habitats for botanists to explore.

8 Northwest

The Northwest is a collage of ancient to recent, exotic to local, terranes from around the globe that have been accreted to the North American continent from the Paleozoic era to the present. Comparable terranes have been accreted elsewhere around the North American craton, but they are most prevalent in the Northwest, from the Northern Cascade Mountains and the Okanagan Highland to the Bering Sea.

Ultramafic rocks are common from the Brooks Range through Alaska, the Yukon Territory, and British Columbia to Washington. They are in fragments of oceanic crust in accreted terranes between 48 and 68°N latitude (Fig. 8-1), and they are in Alaskan-type concentric bodies in Southeast Alaska (Himmelberg and Loney 1995) and in British Columbia (Nixon et al. 1997).

Much of northwestern North America is mountainous. The ultramafic geoecosystems are mostly on moderately steep to very steep hills and mountains. Climates are mostly very cold to cold, but cool on the San Juan Islands. They range from very wet in Southeast Alaska to semiarid at the lower elevations east of the Cascade Mountains. Summers are drier than winters from the latitude of the Cascade Mountains southward. Permafrost is common in areas with ultramafic soils in the watershed of the Yukon River, but no ultramafic soils were found to be frozen through summers.

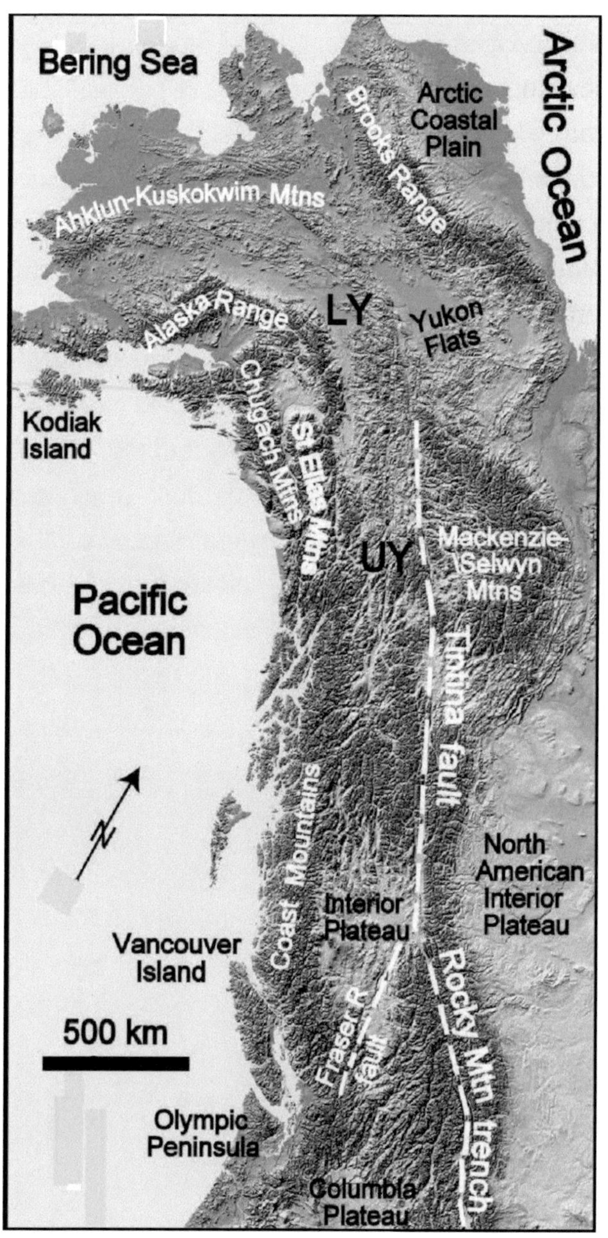

Figure 8-1. Physiography of the Northwest. LY, lower Yukon, UY, upper Yukon.

A. **Geology**

The ultramafic rocks in northwestern North America are in, or among, a complex collage of autochthonous or parautochthonous (definition in *Glossary*) terranes, magmatic arcs, back-arc basins, and allochthonous (exotic) oceanic terranes. They have been scrambled by faulting and by hundreds of kilometers of displacement along mostly dextral (right-lateral) strike-slip faults (Orr and Orr 1996, Colpron et al. 2007, Nelson et al. 2013).

Geologic units in British Columbia and the Yukon are in five belts separated by major faults (Mathews and Monger 2005, Cannings et al. 2011). The belts, from east to west, are the Foreland (deformed terrain in Fig. 8-2), Omineca, Intermontane, Coast, and Insular belts.

During the middle of the Paleozoic, rifting along the margin of Laurentia, the precursor of North America, opened basins that filled with seawater and oceanic sediments, including limestone. Meanwhile, exotic Neoproterozoic to Paleozoic terranes coalesced far offshore in the Panthalassa (worldwide) Ocean. Oceanic crust developed in basins among detachments of rifted continental crust and magmatic arcs of the peri-Laurentian Omineca and parautochthonous terranes of Stikinia and Quesnellia. Remnants of oceanic crust from these terranes are in the Slide Mountain, Cache Creek, Bridge River, Windy-McKinley, and Seventymile (YT unit, Fig. 8-2) terranes, and in the Chulitna terrane along the Talkeetna fault near its junction with the Denali fault at Broad Pass in the Alaska Range. The largest of these, the exotic Cache Creek-Bridge River terrane, was formed in short-lived oceanic intra-arc basins near Laurentia during the Permian (late Paleozoic). Terranes of the Omineca, Intermontane, and Insular belts were attached to one another and to the North American continent during the middle of the Mesozoic era. The Coast belt is composed mainly of Mesozoic to Paleogene oceanic and continental granitic rocks. The Insular Belt contains the exotic Wrangellia and Peninsular terranes.

The Arctic Alaska unit and the Seward and Ruby terranes are Neoproterozoic to Paleozoic pericratonic terranes derived from Baltica, Siberia, and/or Laurentia and eventually attached to North America (Amato et al. 2009). Ultramafic rocks are common in the Mesozoic oceanic Angayucham, Tozitna, Innoko, and Livengood terranes exposed along the margins of the southern limb of the Ruby geanticline or parallel to it, and in the Angayucham terrane where it is associated with the northern limb of the Ruby terrane along the southern margin of the Arctic Alaska unit (Fig. 8-2, Patton et al. 1994, Harris 1998, Colpron et al. 2007). The Koyukuk, Kahiltna, and Togiak terranes in western Alaska are Mesozoic magmatic arc terranes, and the Goodnews terrane is an oceanic associate of the Togiak terrane.

The Chugach terrane consists of very thick late Cretaceous to Paleogene subduction trench deposits, mostly turbidites. It contains some ophiolite. Ultramafic rocks along the Border Range fault, from Kodiak Island around the Chugach Mountains and southward as far as Baranof Island, contain more calcic pyroxenes than might be expected from the peridotites of ophiolites and appear to be cumulates from the fractionation of basalt magma (Burns 1985).

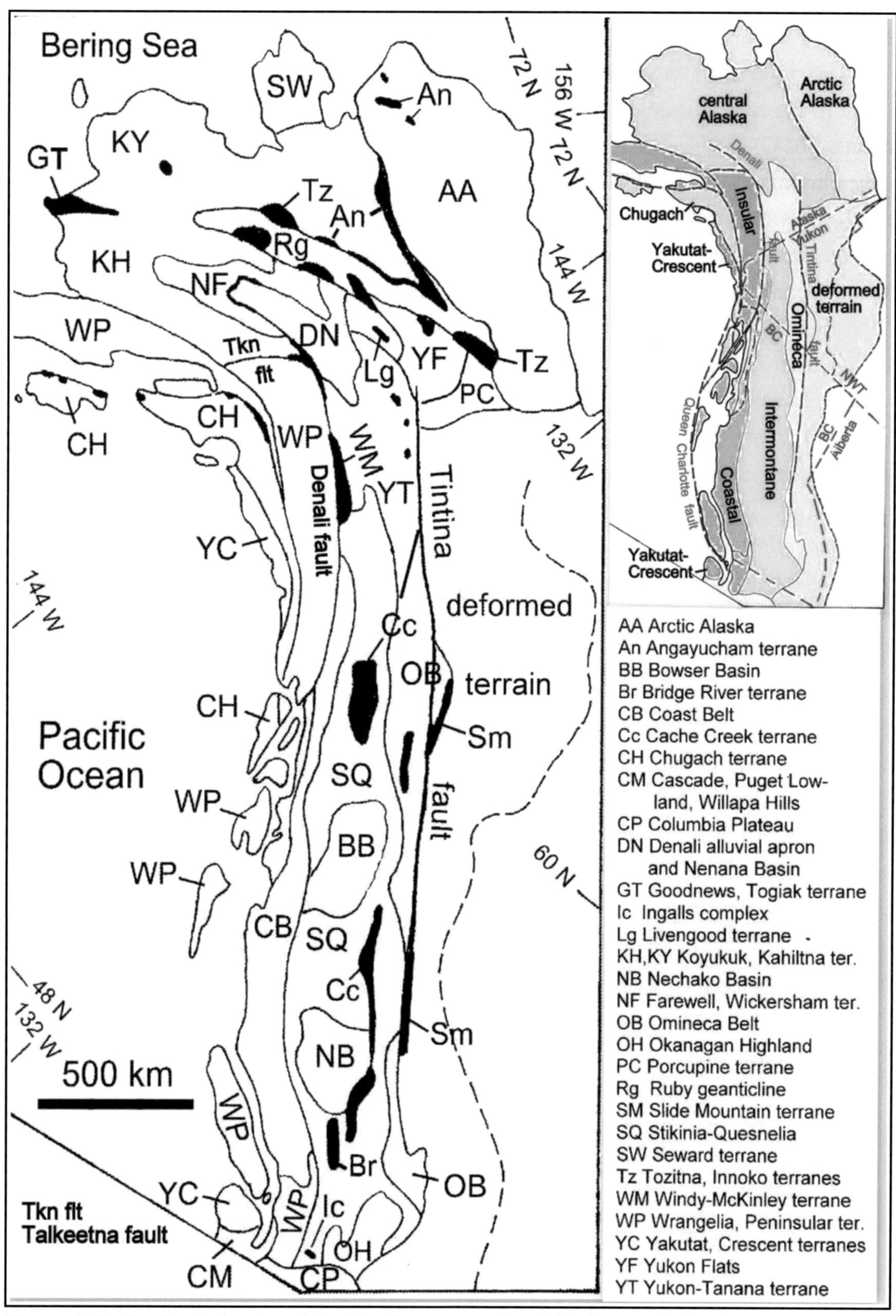

Figure 8-2. Geological units of the Northwest, north of 48 degrees latitude.

The Crescent (or Siletzia) and Yakutat are Cenozoic terranes (YC *in* Fig. 8-2). Recent movement of the Yakutat terrane may be responsible for uplift in the Chugach and Saint Elias Mountains. The CM unit (Fig. 8-2) consists of Cenozoic volcanics of the Cascade Mountains, Cenozoic plutonic and sedimentary rocks of the Willapa Hills, and Quaternary deposits in the Puget Lowlands. There are small exposures of ultramafic rocks in the Okanagan Highlands (unit OH *in* Fig. 8-2) that may belong to the Slide Mountain terrane (Mathews and Monger 2005).

There are many Alaskan-type concentrically zoned bodies in Wrangellia (for example on Duke Island, Irvine 1974) and in the Intermontane Belt (Nixon et al. 1997). All of these Alaskan-type bodies contain hornblende and clinopyroxene rich ultramafic rocks, and dunite is exposed in the centers of some of them.

B. Physiography and Climate

Massive east-west mountain ranges, the Brooks and Alaska Ranges (Fig. 8-1), divide much of Alaska into three temperature zones. Permafrost is continuous in the arctic area north of the Brooks Range, discontinuous in central Alaska, from the Brooks Range to the Alaska Range, and only sporadic south of the Alaska Range (Ferrians 1965). Ultramafics occur as far as 68.5°N in the Brooks Range, but the soils there are expected to be stony and lack permafrost.

Central Alaska is a plateau and plains region with many areas of low mountains (Wahrhaftig 1966). It is represented climatically by Fairbanks (Fig. 8-3), which is very cold. The mean annual precipitation in Fairbanks is only 275 mm, but much of that is summer rainfall, and the summers are short. Consequently, there are no dry seasons. The topography and climate are similar in much of the upper Yukon and

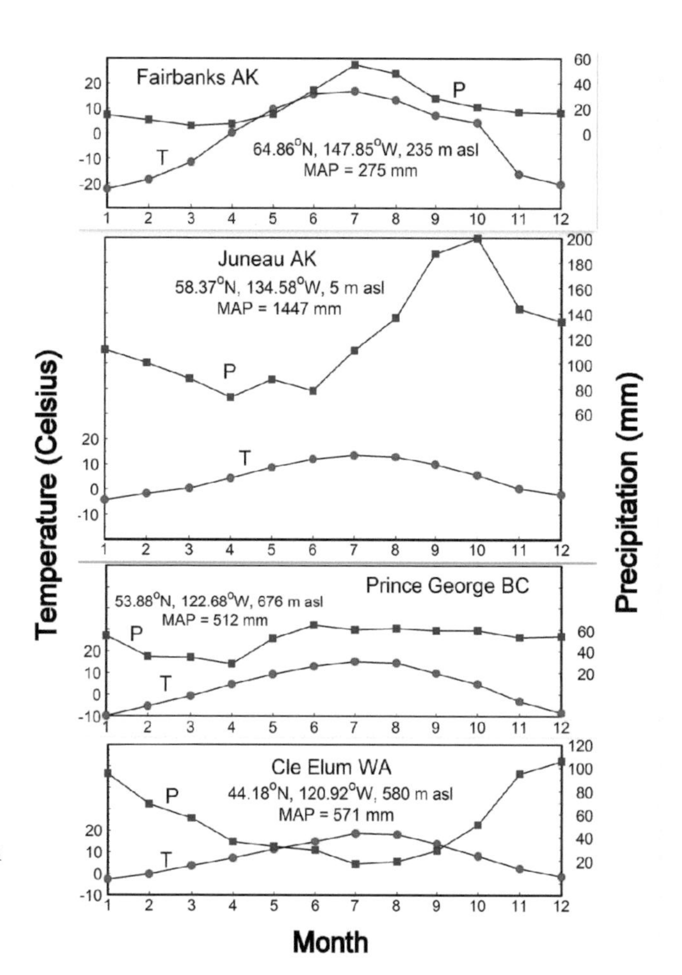

Figure 8-3. Mean monthly temperatures and precipitation at Cle Elum WA, Prince George BC, Fairbanks AK, and Juneau AK.

southward west of the Pelly Mountains to the high Stikine Plateau, which is west of the Cassiar Mountains. There is permafrost in nonultramafic soils as far south as the Skeena and Omineca Mountains, which are north of the Interior Plateau of British Columbia.

Denali (Mt. McKinley, 6168 m) in the Alaska Range is the highest peak in the USA and Mt. Logan (5959 m) in the Saint Elias Mountains is the highest in Canada.

Anchorage, south of the Alaska Range, receives about 430 mm of precipitation annually, with the greatest amounts from late in the summer into the autumn. Precipitation is much greater from Southeast Alaska, as represented by Juneau (Fig. 8-3), southward along the Pacific coast. Peak discharges from the streams of this coastal area are during the summer when the mountain glaciers are melting.

Much water is extracted from the prevailing westerly winds as they cross the Coast Mountains of British Columbia, depriving the Interior Plateau of British Columbia and the Okanagan Highlands of moisture. Nevertheless there are no pronounced dry seasons on the Interior Plateau, because of the short summers (Prince George, Fig. 8-3). The prevailing westerly winds move southward during winters and back northward during summers. Thus areas on the east side of the Cascade Mountains have summer dry seasons (Cle Elum, Fig. 8-3). Many soils east of the crest of the mountains, such as those in the Ingalls complex have xeric (summer dry) soil moisture regimes.

C. Lower and Upper Yukon

The soils and vegetation of four areas were chosen to represent the ultramafic landscapes in the Yukon drainage area. These areas are Caribou Mountain on the margin of the Brooks Range, the Livengood terrane and American Creek in the lower Yukon, and the Atlin-Surprise Lake area. in the upper Yukon.

8C1 Caribou Mountain

Caribou Mountain (970 m asl) is near the head of the Kanuti River, northwest of the Yukon Flats (Fig. 8-1). It consists of serpentinized peridotite and gabbro in the Kanuti segment of the Angayucham terrane (Patton et al. 1994, Fig. 8-2). Several holes were dug in ultramafic soils across the northeast end (700 m asl, 66.385°N, 12 km south of the Arctic Circle, and 150.64°W) of the mountain in search of permafrost. No permafrost was found on the mountain, but ice was found within a meter depth on a granitic pediment adjacent to Caribou Mountain during August.

A view from the northeastern summit toward the main summit of Caribou Mountain (Fig. 8-4A) shows alpine tundra on the upper slopes (Fig. 8-4B) and semidense to open white spruce forests on the lower slopes. Plants on lithic Entisols of the summit are predominantly eightpetal mountain-avens (*Dryas octopetala*) and cushion silene (*Silene acaulis*), with lichens on the rocks; arctic sandwort (*Minuartia arctica*) and arctic lupine (*Lupinus arcticus*) are common and polar grass (*Arctagrostis* sp) is present.

Soils on the upper mountain sideslopes are commonly very cold Inceptisols (Haplocryepts, Fig. 8-4B) with a dense mat of alpine tundra vegetation. Stepped slopes and stone stripes are common. The common plants are crowberry (*Empetrum nigrum*), alpine blueberry (*Vaccinium uliginosum*), a dwarf willow (*Salix* spp.), Labrador tea (*Ledum palustre* ssp. *decumbens*), mountain-avens (*Dryas* sp), arctic mountain heather (*Cassiope tetragona*), and mosses, and lichens (especially *Cladina* spp.) are common.

Figure 8-4. Landscapes of Caribou Mountain. A. A view westward from the eastern summit. B. Upper sideslope. C. A nivation hollow on a sideslope. D. Mountain footslope.

Nivation hollows with wet Inceptisols (Histic Cryaquepts) and sorted stones are present on the lower northeast mountain sideslopes (Fig. 8-4C). The main plants in the hollows are shrub birch (*Betula nana*), Labrador tea, arctic mountain heather, alpine blueberry, grasses, and sedges. A common sedge is *Carex bigelowii* and sphagnum mosses are common.

Vegetation in the open forests on deep Inceptisols of the mountain footslopes (Fig. 8-4D) is mostly white spruce (*Picea glauca*), alder (*Alnus viridis* ssp *crispa*), crowberry, alpine blueberry, Labrador tea, mosses, and lichens (mostly *Cladina* sp. and *Flavocetraria* sp). Arctic mountain heather, shrub birch (*Betula nana*), and fescue (*Festuca altaica*) and other grasses are common. Shrub birch is more common in the transition to alpine tundra.

A frozen soil (a Historthel) on a granitic pediment no more than a kilometer southeast from Caribou Mountain is shown in Figure 8-5. As described by Alexander (2004) it has Oi, Oe, A, Bw, Ab, and Cf horizons. The O (organic matter) horizon is 38 cm thick and the frozen (Cf) horizon is 21 cm below the O-horizon. The petroleum pipeline visible in the background is above ground where there is permafrost and below ground where there is none. The plant community was characterized by sparse black spruce, common shrubs, sparse forbs, common sedges and rushes, plentiful mosses, especially sphagnum, and abundant lichens. The common shrubs were Labrador tea, alpine blueberry, dwarf birch, mountain alder, and lingonberry. Along with abundant *Cladina* spp., there was some curled snow lichen (*Flavocetraria cucullata*). This is near the northern limit of black spruce; white spruce is the only spruce nearby on Caribou Mountain.

Figure 8-5. A nonultramafic Gelisol on a granitic pediment near Caribou Mountain.

8C2 Livengood

Loney and Himmelberg (1988) reported on the geology of the Livengood terrane (Lg in Fig. 8-2). Ultramafic soils were described at three Livengood sites in low mountainous terrain at about 65.5°N and 148.5°W (Table 8.1): Livengood site 1 (LG3) on a broad ridge at 550 m asl, site 2 (LG2) on a steep south-facing slope, and site 3 (not in Table 8.1) on a very steep north-facing slope. Soils on the ridge and south-facing slope are moderately deep and deep Inceptisols (loamy-skeletal, and loamy-skeletal over fragmental, magnesic Typic Haplocryepts) on frost-shattered serpentinized peridotite bedrock with about 0.2 m of loess incorporated into the surface horizons. The soil on the north-facing slope is a very deep, wet Inceptisol (loamy-skeletal, magnesic Oxyaquic Haplocryept) in serpentinized peridotite colluvium with about 0.25 m of nonultramafic cover. Loess on the Livengood soils was probably deposited thousands of years ago. Low ratios of A horizon exchangeable Ca/Mg <0.5 (Table 8.2) indicate that the loess has been thoroughly mixed into ultramafic materials at Livengood sites 1 and 2. The surface soil pH of the ridge soil (site 1, LG3) is low (pH 5.8) for a ultramafic soil, and has KCl extractable acidity >0.10 mmol/kg which is unusually high for an ultramafic soil.

Vegetation on the Inceptisol of the south-facing slope was an open white spruce (*Picea glauca*)/kinnikinnick (*Arctostaphylos uva-ursi*)–low juniper (*J. communis*)/arctic sandwort (*Minuartia arctica*) plant community. Vegetation on the Inceptisol of the ridge was a paper birch (*Betula papifera*)–black spruce (*Picea mariana*)/Labrador tea (*L. groenlandicum*)–lingonberry (*Vaccinium vitis-idaea*)–alpine blueberry (*Vaccinium uliginosum*)/bluejoint reedgrass (*Calamagrostis canadensis*)–fescue grass (*Festuca altaica*) plant community. Vegetation on the wet Inceptisol of the north-facing slope was a black spruce–paper birch/Labrador tea–alpine blueberry/sphagnum plant community. The ultramafic soil on the south-facing, but not the one on the north-facing slope, had a plant community that was distinctly different from those on adjacent nonultramafic soils. The common plant community on steep nonultramafic south-facing slopes was a relatively dense white spruce (*Picea glauca*)–paper birch/lingonberry/northern comandra (*Geocaulon lividum*) woodland. There was no northern comandra on the ultramafic soils, nor any juniper on adjacent nonultramafic soils. The common lichens on the well-drained ultramafic soils were crinkled snow lichen (*Flavocentraria nivalis*) and antlered powderhorn (*Cladonia sublata*) on the south-facing slope, and antlered powderhorn, curled snow lichen (*Flavocetraria cucullata*), thorn lichen (*Cladonia uncialis*), and sieve lichen (*Cladonia multiformis*), and lesser amounts of studded leather-lichen (*Peltigera aphthosa*) on the broad ridge.

Plants have been inventoried in the Ultramafic Slide Research Natural Area, near Livengood. The rock types in the RNA are clastic sedimentary rocks, dolomite, chert, greenstone, and serpentinite. Much of the serpentinite has been exposed and is maintained in barren condition by erosion. The serpentinite is in the 500 to 900 m altitude range. Juday (1992) found that the plant species with greatest tolerance of the serpentine conditions were thoroughwax (*Bupleurum americanum*) and arctic sandwort (*Minuartia arctica*), and that sorrel (*Rumex acetosa*) grows on the serpentine barrens. Only a crustose lichen inhabits the most severe barrens.

8C3 American Creek, Yukon Plateau

Soils and vegetation were observed on serpentinized harzburgite along the Taylor Highway where American Creek cuts through ultramafic rocks of the Seventymile terrane (Patton et al. 1994) just south of the village of Eagle, which is on the Yukon River. Because of very steep slopes, soils in the canyon are mostly either shallow to bedrock or very deep in colluvium. West of American Creek south-facing slopes have open forest and north-facing slopes have semidense forest cover.

Two ultramafic soils were described on west-facing slopes on the east side of American Creek: a lithic, haplic Mollisol (Lithic Haplocryoll) with an open white spruce (*Picea glauca*)–paper birch (*Betula papifera*)/mountain alder (*Alnus viridis* ssp. *crispa*)–alpine blueberry (*Vaccinium uliginosum*)/moss plant community and a very deep, haplic Mollisol (Pachic Haplocryoll) with a semidense paper birch–white spruce/mountain alder/purple reedgrass (*Calamagrostis purpurascens*)/moss plant community. Forbs and sedges were sparse on both soils. Shrubby cinquefoil (*Dasiphora fruticosa*) was widely distributed on adjacent landslides and ultramafic roadcuts, but it did not occur on moss-covered soils. Silt loam textures in the surface of the shallow Mollisol indicate that some loess from the floodplain of the Yukon River, about 12 km to the north of the lithic Mollisol, has reached the site. The loess influx is more obvious in the very deep Mollisol, which is only 8 km from the Yukon River. An E-horizon has formed in the extremely stony surface of the very deep Mollisol.

8C4 Atlin-Surprise Lake

Nahlin ultramafics in the Atlin terrane of the Cache Creek group (Monger 1977) are exposed in the areas southeast and northeast of Atlin, British Columbia, and north of Surprise Lake (Aitken 1959). This is a highly dissected area in the Teslin section of the Yukon Plateau, near the Stikine Plateau (Holland 1976), or perhaps more appropriately, because of strong topographic relief, the Yukon-Stikine Highland (Bostock 1948, Smith et al. 2004). Ultramafic soils and vegetation were observed on an unnamed mountain (Fig. 8-6) north of Surprise Lake and southeast of Mt. Barnham (2087 m asl), about 15 to 18 km northeast of Atlin.

Figure 8-6. Alpine tundra on an unnamed mountain east of Atlin.

Soils and vegetation were observed on southwest-facing slopes from about 1250 m asl near Ruby Creek up to about 1850 m on the summit (Fig. 8-6) of the unnamed mountain, The soils are very shallow to very deep Mollisols (Cryolls) with some rock outcrop and much barren talus, although the talus is far from barren, considering the abundant lichens on the rocks. The upper mountain slopes are covered by a carpet of cespitose willows (*Salix* spp.) and yellow mountain-avens (*Dryas drummondii*) with sedges and grasses, including arctic fescue (*F. altaica*), and patches of mountain heather (*Cassiope tetragona*), cushion silene (*Silene acaulis*), and stonecrop (*Sedum lanceolatum*). The alpine tundra grades downward to arctic fescue meadow with crowberry (*Empetrum nigrum*) and cespitose willows, and with patches of erect willow shrubs (*Salix* sp) and birch (*Betula nana*) and scattered low juniper (*J. communis*) and subalpine fir (*Abies lasiocarpa*). Among the more common serpentine forbs in the alpine meadow were lupine (*Lupinus* sp), burnet (*Sanguisorba canadensis*), fireweed (*Camerion angustifolium*), ragwort (*Senecio lugens*?), and monkshood (*Aconitum delphinifolium*). Further downward, the fir trees form a discontinuous forest around the fringe of Ruby Valley, but there is no forest below about 1300 m in Ruby Valley. The vegetation below 1300 m in the Valley, which has a filling of glacial drift with mixed lithologies, is willow (*Salix* sp)–birch (*Betula nana*) scrub with a soft ground cover of mosses.

D. Gulf of Alaska

There are many exposures of layered gabbro and ultramafic rocks along the Border Range fault from the northeast side of Kodiak Island (Fig. 8-1 and 8-2) across the Kenai Peninsula, around the northwestern end of the Chugach Mountains, and down along the northeastern side of the Chugach Mountains to the Fairweather fault. These layered bodies have been considered to be the roots of magmatic arcs (Burns 1985). Soils and vegetation were described near Eklutna, about 40 km northeast of Anchorage. East of the Chatham Straight in Southeast Alaska, there are several Alaskan-type concentric bodies. Soils and vegetation were described on one of those concentric bodies on the Cleveland Peninsula.

8D1 Eklutna

A small body of weakly serpentinized, layered wherlite, dunite, and pyroxenite (Patton et al. 1994), about 40 km northeast of Anchorage, is exposed from near sea level adjacent to the mouth of the Knik River over a ridge to the valley of the Eklutna River. Mount Eklutna on the ridge is 1253 m asl. The mean annual precipitation is about 400 mm.

North of the Eklutna River, soils and vegetation were observed across a ridge that extends northwest from West Twin Peak. The typical bedrock is clinopyoxenite with a high specific gravity (SG=3.2) that confirms the visual impression that alteration of it has been negligible. Slopes are steep on the north and very steep to steep on the southwest side of the ridge. The soils are Inceptisols (Haplocryepts) and Mollisols (Haplocryolls). A forest of predominantly white spruce (*Picea glauca*) and paper birch (*Betula papyrifera*) prevails on the north-facing slope, with much bunchberry (*Cornus canadensis*) and mosses in the understory. The southwest facing slopes have shallow (lithic) soils on very steep inclines and deeper soils on steep inclines. The vegetation on the very steep slopes is grasses, largely alpine fescue, and kinnikinnick (*Arctostaphylos uva-ursi*), prickly saxifrage (*S. tricuspidata*), low juniper (*J. communis*), buffaloberry (*Shepherdia canadensis*), and saskatoon (*Amelanchier alnifolia*), with sparse lichen (*Cladina* sp). Semidense to open forest prevails on the steep southwest-facing slopes. It is predominantly aspen (*Populus tremuloides*) and paper birch (*Betula payrifera*), with much buffaloberry, rose (*R. acicularis*), and with fireweed (*Chamerion angustifolium*) and other forbs in the understory, and with sparse white spruce and mosses. Balsam poplar (*Populus balsamifera*) is present in forests of the lower slopes on both north and southwest sides of the ridge.

8D2 Golden Mountain

Ultramafic soils and vegetation were described on Golden Mountain (Fig. 8-7), which is near the western end of the Cleveland Peninsula, about 35 km northwest of Ketchikan, Alaska. The ultramafic rocks are in a concentrically zoned body (Ruckmick and Noble 1959). The dunite core of the body is centered on Golden Mountain, which has a maximum elevation of 773 m, and the outer zones are at lower elevations down practically to sea level. Golden Mountain is five or six kilometers east of Mt. Burnett, at the opposite end of a ridge that joins them (topographic map in Alexander et al. 1989, 1994a). The older literature places Mt. Burnett at the current location of Golden Mountain. Glaciers from the east covered the area during the last ice age, but any glacial drift that might have covered the summit of Golden Mountain has been washed away. Currently the mean annual precipitation is about 2000 mm.

Alexander et al. (1989) described an Inceptisol on dunite at 740 m asl on Golden Mountain and at 420 m asl a Spodosol in ultramafic glacial till with a thin cover of nonultramafic material. The Inceptisol had an Oe-A-Bw-C-R horizon sequence with a friable, brown (7.5YR 4/5, moist), slightly acid Bw horizon containing 13% (126 g/kg) citrate-dithionite extractable Fe (representative of 180 g/kg FeOOH). It is an Inceptisol (Typic Haplocryept).

Figure 8-7. Golden Mountain, a view from east to west across the Cleveland Peninsula

The ultramafic Spodosol described by Alexander et al. (1989, 1994a) had an A-E-2Bs-2C-2Cqm soil horizon sequence, with the A and E horizons in nonultramafic material. It is an ephemerally wet Spodosol (Oxyaquic Duricryod). Although there are a few strong brown (7.5YR 4/6) mottles in the albic E horizon, these sparse redoximorphic features were not considered to be adequate for a more continuously wet Spodosol (Aquod). A nonultramafic Spodosol on glacial till near the ultramafic Spodosol was appreciably wetter and was definitely a wet Spodosol (Cryaquod). Acid oxalate extractable Fe was only 43 g/kg from the Bhs horizon of the nonultramafic Spodosol compared to 82 g/kg from the 2Bs horizon of the ultramafic Spodosol, which lacked a Bhs horizon. Silica cementation in glacial till beneath the ultramafic Spodosol was verified by laboratory investigations (Alexander et al. 1994b).

The nonultramafic Spodosol on Golden Mountain supports a Sitka spruce–hemlock–western red cedar–yellow cedar forest, with red alder and many species of shrubs in the understory. Douglas-fir does not reach this far north in the Alaska Coast Ranges. The ultramafic soils have more open plant communities, with sparse shore pine (*P. contorta* var *contorta*) trees. Also, there are some yellow cedar (*Callitropsis nootkatensis*) and mountain hemlock (*Tsuga mertensiana*) trees on the ultramafic Spodosol, but no western hemlock (*Tsuga heterophylla*), which occurs on the nonultramafic Spodosol. The ultramafic Spodosol was sampled on a secondary ridge and the nonultramafic soil was just below the ridge--the slightly more protected aspect of the nonultramafic Spodosol may, or may not, account for some of the differences in plant communities, but the major differences are judged to be related to substrate differences. Shrubs on the ultramafic Inceptisol were western moss heather (*Cassiope mertensiana*), crowberry (*Empetrum nigrum*), yellow mountain heath (*Phyllodoce glanduliflora*), and bog blueberry (*Vaccinum uliginosum*), and those on the ultramafic Spodosol were crowberry, low juniper (*J. communis*), and Labrador tea *(Ledum glandulosum*). Several species of shrubs that occurred on the nonultramafic Spodosol and are common in the area were not found on the ultramafic soils. None of forbs or grasses are endemic species (personal communication, P. Krosse 1989).

E. Interior British Columbia

Most of the ultramafic rocks in British Columbia are in the Fraser River drainage basin, where there are numerous exposures of ultramafic rocks in oceanic terranes and in Alaskan-type concentric bodies. Soils were sampled at several sites in the Shulaps Range and in the Tulameen area (Bulmer et al.1992, Bulmer and Lavkulich 1994). Subsequently, I have described the geoecological landscapes in those same areas.

Paul Sanborn and his collaborators (Sanborn 2013) have sampled and analyzed a number of very cold soils across the Cache Creek terrane on the Interior Plateau, near Prince George. Most of the soils had glacial, colluvial, or aeolian deposits on them. The soils that were in predominantly ultramafic materials were forested Entisols and Inceptisols and subalpine Mollisols. They were generally slightly acid to neutral in their subsoils, even with strongly acid LF (plant litter) horizons. Depths to slightly acid or neutral pH was deeper in soils with extremely acid H (humus) horizons than in soils without those horizons. Exchangeable Al was very low (<0.001 meq/100 g, or <0.01 mmol$_{+}$/kg) in slightly acid to neutral ultramafic soil horizons, and total Al in them was generally <3% (or <6% Al-oxide). Subsoil (B and C horizons) molar exchangeable Ca:Mg ratios were in the 0.05 to 0.35 range. Nickel at about 0.2% was high enough to be potentially toxic to many plants in acid soils.

8E1 Shulaps Range

Ultramafic bodies of the similar or equivalent Cache Creek and Bridge River terranes (Fig. 8-2) occur along faults that branch off from the Fraser River fault and extend northwestward into the Coast Mountains and southeastward across the Interior Plateau of British Columbia. Prominent among the ultramafic bodies are the large Shulaps body along the Yalakom fault and the Coquihalla ultramafics along the Hozameen fault, and there is a relatively small Pioneer peridotite near Bralorne, southwest of the Shulaps Range (Monger1977, Wright et al. 1982). Apparently, the Shulaps body and the Coquihalla ultramafics are representatives of what was once one body before they were separated by about 80 km of displacement along the Fraser River fault (Mathews and Monger 2005). Bulmer and Lavkulich (1994) have described and characterized soils in the Shulaps Range and for an Inceptisol in glacial till of the Coquihalla.

The Shulaps ultramafic body has been thrust southwestward over granitic plutons of the Coast Mountains. It is about 30 km by 15 km and is one of the largest ultramafic bodies in British Columbia (Wright et al. 1982). Serpentinized peridotite of the body is exposed on the east side of the Shulaps Range down to the Yalakom River, which is a branch of the Bridge River. The Bridge River flows eastward from the Coast Mountains into the Fraser River. A cover of fine pumice about 20 cm thick was spread over soils of the Shulaps body about 2350 years ago (Mathewes and Westgate 1980) and has been redistributed by wind and slope wash.

Five soils were described and sampled from near the upper limit of ultramafic rock down to glaciated terrain along the Yalakom River (Tables 8.1 and 8.2). The highest site, at 2450 m asl (Shulaps 1) had a rudimentary soil (a Cryorthent) with a surface pavement of loose gravel and plant cover of common moss campion (*Silene acaulis*), and sparse grasses. The surface layers of the soils at the Shulaps 2, 3, and 4 sites were in tephra. and the soils had A, E, or Bs horizons, or all of these horizons. The upper Bs horizons of the Shulaps 2 and 4 soils had more than 2% (20 g/kg) citrate-dithionite extractable Fe, but barely any Al, and insufficient acid oxalate extractable Fe to surpass the 0.5% Al + Fe/2 requirement for spodic horizons (Bulmer and Lavkulich 1994, Soil Survey Staff 1999). Thus, the soils at Shulaps 2, 3, and 4 sites were Inceptisols. The Shulaps site 3 had a semidense stand of lodgepole pine with an understory of common kinnikinnick (*A. uva-ursi*) and low juniper (*J. communis*), sparse Engelmann spruce, Cascade azalea (*Rhododendron albiflorum*), and subalpine fir, and traces of *Lilium philadelphicum* var *andinum* and other herbs. The lowest site, at 1370 m asl (Shulaps 5) was a shallow Alfisol in glacial till over serpentinized peridotite. It was an Alfisol with plentiful Douglas-fir and some Rocky Mountain juniper (*J. scopulorum*), common low juniper, kinnikinnick, and russet buffaloberry (*Shepherdia canadensis*), sparse saskatoon, spiraea (*S. betulifolia*), barberry (*Mahonia repens*), and rose (*Rosa* sp), plentiful wheatgrass (*P. spicata*) and purple reedgrass (*Calamagrostis purpurascens*), and traces of nodding onion (*Allium cernuum*) and yarrow (*Achillea millefolium*).

A more general assessment of the serpentine vegetation on the Shulaps body might be open Douglas-fir forests on south-facing slopes at low elevations and dense forest from there up to the subalpine area that is dominated by sparse whitebark pine and low shrubs. The understory in the open Douglas-fir (*Pseudotsuga menziesii*) forest is commonly dominated by wheatgrass (*Pseudoroegneria spicata*), low juniper (*J. communis*), kinnikinnick (*Arctostaphylos uva-ursi*), buffaloberry (*Shepherdia canadensis*), stonecrop (*Sedum lanceolatum*), and moss. Dense forests are predominantly subalpine fir (*Abies lasiocarpa*) and lodgepole pine (*Pinus contorta*) grading to whitebark pine near the upper elevation limit of forest. The main forest shrubs are white (or Cascade) rhododendron *(Rhododendron albiflorum*) and black huckleberry (*Vaccinium membranaceum*) on north-facing slopes at lower elevations and otherwise low juniper (*J. communis*) with common kinnikinnick (*Arctostaphylos uva-ursi*) and buffalo berry (*Shepherdia canadensis*). Low juniper, dwarf willow (*Salix* sp) and kinnikinnick are the main subalpine shrubs and buckwheat (*Eriogonum* sp) is a common subshrub. Some of the main subalpine forbs are stonecrop (*Sedum lanceolatum*), lupine (*Lupinus* sp), spotted saxifrage (*Saxifraga bronchialis*), paintbrush (*Castilleja* sp.), and sandwort (*Minuartia* sp.).

Kruckeberg (1969) has described vegetation of the Pioneer peridotite, which he refers to as the Bralorne and Choate area. He noted dense forests of *Tsuga mertensiana–Callitropsis nootkatensis* on nonultramafic sites around the ultramafics. He described openings with Sitka sedge (*Carex aquatilis*) in seeps on pyroxenite, and on pyroxenite talus he described a *Rubus leucodermis/ Adiantum aleuticum–Aspidotis densa* plant community.

8E2 Tulameen complex

The Tulameen complex is considered to be an Alaskan-type concentric body (Nixon et al. 1997). It has a dunite core about 4 by 1 km (Findlay 1969). The Tulameen complex is exposed mostly on north-facing slopes of Olivine Mountain, which is south of the Tulameen River, and on south-facing slopes of Grasshopper Mountain, which is north of the River. Glacial drift of mixed lithologies covers the lower mountain slopes, along the Tulameen River, and is present in patches all of the way up the mountains. It has been completely removed from very steep and most steep slopes.

Inceptisols are common on moderate to steep slopes and they are very shallow to very deep. Extremely stony Entisols are common on very steep slopes. There are some ultramafic Mollisols, which appear to be associated with pyroxenite, more than with the dunite core of the Tulameen complex, but observations were insufficient to verify soil parent material related differences. There are soil site data in Tables 8.1 and 8.2. The vegetation on very steep slopes of Grasshopper Mountain was dominated by sparse Douglas-fir, low juniper (*J. communis*), and bluebunch wheatgrass (*P. spicata*). These same species occur on moderately steep and steep Inceptisols, with denser Douglas-fir, sparse ponderosa pine, common kinnikinnick (*A. uva-ursi*), and some saskatoon (*A. alnifolia*), roses (*Rosa* spp.), chokecherry (*Prunus virginiana*), prostrate buckwheats (*Eriogonum* spp.), mountain sandwort (*Arenaria capillaris*), serpentine fern (*Aspidotis densa*), and sparse yew (*Taxus brevifolia*). Subalpine fir is common on upper north-facing slopes, and there is some western white pine (*P. monticola*).

Bulmer and Lavkulich (1994) sampled ultramafic soils at 1260 m asl on Grasshopper and 1730 m asl on Olivine Mountain (Fig. 8-8). Only small amounts of clay have accumulated in these soils. Mineralogical analyses showed the very fine sand fraction of the soil from Olivine Mountain to be olivine and the clay fraction was dominated by serpentine.

Kruckeberg (1969) reported that there is an impressive array of serpentine effects on the vegetation in the Tulameen area. He noted the xeric nature of the serpentine vegetation relative to the surrounding vegetation. The widespread serpentine "faithful" ferns *Aspidotis densa*, *Polystichum kruckebergii*, and *P. lemmonii* wee common. He sampled several relevés from the vicinity and described the vegetation in open woodlands there as *Pseudotsuga menziesii/Juniperus communis–Prunus virginiana* var *demissa/Aspidotis densa–Polystichum kruckebergii* associations.

Figure 8-8. A view from Grasshopper Mountain across the Tulameen River to Olivine Mountain.

A detailed floristic and ecological analysis of the Grasshopper Mountain area was done by Lewis and Bradfield (2003). They sampled 71 plots: 26 on ultramafic soils, 35 on nonultramafic soils, and 10 on glacial till soils in terrains of similar topography. Species diversity was not significantly different (Shannon or Simpson diversity indices) between ultramafic and nonultramafic sites. This conclusion differs from general statements made by Kruckeberg (1969), Brooks (1987), and others on the overall reduction of species diversity on ultramafic substrates. Of all species recorded 28% (49 taxa) were found only on ultramafics. The majority of these species are not true endemics but are species that are out of their normal distribution ranges, as a result of the unique ultramafic substrate. They are local serpentine indicators. Plants in several families including Apiaceae, Caryophyllaceae, Poaceae, and fern families (Pteridophytes) are more common on ultramafics than off, while those in the Liliaceae, Rosaceae, Ranunculaceae, Betulaceae, Capriophyllaceae, Grossulariaceae, and Salicaceae are more common off ultramafics, locally. Thirty-five species were identified as local indicators of serpentine by Lewis and Bradfield (2003).

Table 8.1 Ultramafic soil site and vegetation data. Sites are ordered in a geographic sequence from northwest (central Alaska) to south (Northern Cascade Mountains, Washington).

Site[a]	Lat. and Long.	Alt. (m)	Slope	Parent[b] Material	Soil Depth and Order	Precip. (mm/year)
Livengood 1 (Site LG3)	65.51° 148.51°W	550	NW 18%	loess/ serpent.	mod. deep Inceptisol	300
common trees/abundant shrubs/grass/lichens: PIMA-BEPA/LEPA-VAVI/CACA						
Livengood 2 (Site LG2)	65.49°N 148.37°W	400	SW 42%	loess/ serpent.	deep Inceptisol	300
sparse trees/common shrubs/much moss-lichens: PIGL/ARUU-JUCO						
Shulaps 1 (Site BR2)	51.026°N 122.557°W	2450	S 4%	ultramafic (UM) till	deep Entisol	800
sparse grass/common forbs: SIAC/*Poa* sp.						
Shulaps 2 subalpine	51.026°N 122.555°W	2150	SE gentle	tephra/ UM till	deep Inceptisol	770
sparse trees/shrubs/forbs, patches of heath: PIAL/JUCO-SALIX-ARUU/forbs						
Shulaps 3	51.0°N 122.4°W	1750	N 22%	tephra/ serpent.	mod deep Inceptisol	600
abundant trees/common shrubs/common forbs: PICO/ARUU-JUCO-RHAL-AMAL/LUAR						
Shulaps 4	50.9°N 122.3°W	1410	—	tephra/ UM till	deep Inceptisol	440
forest: PSME						
Shulaps 5 (Site BR1)	51.023°N 122.446°W	1370	S 50%	tephra/till/ peridotite	shallow Alfisol	400
very common trees, shrubs/abundant grass: PSME/JUCO-SHCA-ARUU/AGSP-CAPU						
Tulameen 1	49.5°N 120.9°W	1730	—	dunite	shallow Inceptisol	840
open forest/shrubs/grass: PIAL-ABLA/shrubs/grass						
Tulameen 2 (Site GH1)	49.542°N 120.890°W	1480	S 26%	serpent. peridotite	shallow Mollisol	700
sparse trees/common shrubs/very common forbs/grass: PIAL-ABLA/JUCO/forbs/AGSP-POSE						

Site[a]	Lat. and Long.	Alt. (m)	Slope	Parent[b] Material	Soil Depth and Order	Precip. (mm/year)
Tulameen 3 (Site GH2)	49.541°N 120.894°W	1460	NW 58%	serpent. peridotite	mod. deep Inceptisol	700
abundant trees/very common shrubs/common grass: PIAL-ABLA/JUCO-ARUU/AGSP						
Tulameen 4 (Site GH3)	49.54°N 120.9°W	1325	SW 52%	colluvium/ till	deep Mollisol	650
abund. trees/common shrubs/common herbs: PSME/JUOC-PRVI-AMAL/AGSP-LUAR-ARCA						
Tulameen 5	49.54°N 120.9°W	1260	—	ultramafic (UM) till	—	610
open forest/shrubs/grass: PSME-PICO/shrubs/grass						
Ingalls 1 (Site IC3)	47.458°N 120.971°W	1805	W 34%	serpent.	shallow Entisol	2500
sparse trees/common forbs/sparse grasses: ABLA/PHDI–ACMI						
Ingalls 2 (Site IC4)	47.458°N 120.972°W	1790	W 36%	tephra/ serpent.	mod. deep Inceptisol	2500
plentiful trees/plentiful shrubs/sparse forbs: ABLA-TSME-PIAL/VASC						
Ingalls 3 (Site IC2)	47.457°N 120.974°W	1650	SW 72%	serpent.	shallow Inceptisol	2000
sparse trees/common shrubs/common forbs/plentiful grasses: PSME/JUCO//ACMI/AGSP						

[a] Soils at underlined locations were sampled by C. Bulmer. [b] All of the dunite and peridotite were at least partially serpentinized. [c] Soil depths: shallow, 10-50 cm; moderately deep, 50-100 cm, deep >100 cm. Vegetation: crown cover abundance classes are trace <1%, sparse (1-3%), common (3-10%), plentiful (10-30%), abundant (30-60%), and dominant >60%. Plant species abbreviations: ABAM, silver fir (*Abies amabilis*); ABLA, subalpine fir (*Abies lasiocarpa*); ACMI, *Achillea millefolium*; AGSP, wheatgrass (*Pseudoregnaria spicata*); ALCR, mountain alder (*Alnus viridis* ssp *crispa*); AMAL, saskatoon (*Amelanchier alnifolia*); ARCA, *Arenaria capillaris*; ARUU, kinnikinnick (*Arctostaphylos uva-ursi*); BEPA, paper birch (*Betula papyrifera*); CACA, bluejoint reedgrass (*Calamagrostis canadensis*); CAPU, purple reedgrass (*Calamagrostis purpurascens*); CANO, yellow cedar (*Callitropsis nootkatensis*); JUCO, low juniper (*J. communis*); LEPA, Labrador tea (*Ledum palustre*); LUAR, *Lupinus arcticus*; PHDI, *Phlox diffusa*; PIAL, whitebark pine (*P. albicaulis*); PICO, lodgepole pine (*P. contorta*); PIGL, white spruce (*Picea glauca*); PIMA, black spruce (*Picea mariana*); POSE, bluegrass (*Poa secunda*); PRVI, *P. virginiana* var *demissa;* PSME, Douglas-fir (*Pseudotsuga douglasii*); RHAL, *Rhododendron albiflorum*; SALIX, willow (*Salix* spp.); SHCA, buffaloberry (*Shephardia canadensis*); SIAC, moss campion (*Silene albicaulis*); TSHE, hemlock (*Tsuga heterophylla*); TSME, hemlock (*Tsuga mertensiana*); VAME, black huckleberry (*Vaccinium membranaceum*); VASC, grouseberry (*Vaccinium scoparium*); VAUL, alpine blueberry (*Vaccinium uliginosum*); VAVI, lingonberry (*Vaccinium vitis-idaea*).

Eight rare (rare in British Columbia) plant taxa occur on Grasshopper Mountain only on ultramafics. As appears to be the general pattern in northwestern North America, many of these are ferns and include *Aspidotis densa, Polystichum kruckebergii, P. scopulinum,* and *Adiantum aleuticum* (unofficial *A. pedatum* ssp *calderi*). Other plant species considered rare were *Arabis holboellii* var *pinetorum, Crepis atrabarba* ssp *atrabarba, Lupinus arbustus* ssp *pseudoparviflorus,* and *Melica bulbosa.*

Lewis et al. (2004) identified 43 species of bryophytes (mosses and liverworts) in the ultramafic area of the Tulameen complex and an additional 21 species in the surrounding area. They suggested that moss species distributions are more closely related to physical habitat than to rock or soil chemistry.

F. Northern Cascade Mountains

Ultramafic geoecosystems were investigated on Twin Sisters and the Ingalls complex in the Cascade Mountains, and on the eastern San Juan Islands, about 20 or 30 km west of the Cascade Mountains. The Straight Creek fault, an extension of the Fraser River fault, continues through the Northern Cascade Mountains (Fig. 8-1). The Twin Sisters and San Juan Islands are west of the Fraser River–Straight Creek fault and the Ingalls complex (Ic in Fig. 8-2) is east of the fault.

The climate ranges from cool to very cold and from dry to wet. The mean annual precipitation is up to about 2500 mm on Twin Sisters Mountain, about 700 mm on the eastern San Juan Islands, and from nearly 1000 mm up to about 2500 mm on the Ingalls complex, with relatively dry summers (Fig. 8-3, Cle Elum). Precipitation is greater in winter than in summer and snow accumulates and persists through winters in the mountains, but not at the lower elevations nor on the San Juan Islands where snow does not persist throughout winters.

8F1 Twin Sisters

Twin Sisters (Fig. 8-9) comprise the largest exposure of dunite in North America, about 260 km^2 of dunite (Thompson and Robinson 1975). Dunite is named after Dun Mountain in New Zealand. The dominant mineral in the dunite is olivine with minor amounts of pyroxenes and chromite; it lacks serpentinization, except along the faulted boundaries of the dunite.

Figure 8-9. Twin Sisters, with Mt. Baker, a volcano on the left, beyond North Sister.

Glaciers have occupied the Twin Sisters, and one or more glaciers are still present. The higher reaches of the Twin Sisters locality are mostly barren rock and talus lacking soils. There are sparse patches of Inceptisols (Cryepts) and very shallow lithic Entisols (Cryents), however, with stands of sparse mountain hemlock (*Tsuga mertensiana*) and subalpine fir (*Abies lasiocarpa*) and semidense understories of pink mountain heath (*Phyllodoce empetriformis* and western moss heather (*Cassiope mertensiana*) and black huckleberry (*Vaccinium membranaceum*) and sparse patches of very deep Inceptisols with stands of sparse subalpine fir and dense understories of low juniper (*J. communis*). Colluvial footslopes commonly have Inceptisols with semidense stands of mountain hemlock and subalpine fir and understories of pink mountain heath and oval-leaf blueberry (*Vaccinium ovalifolium*), patches of white-flower rhododendron *(Rhododendron albiflorum*), and sparse rusty Menziesia (*Menziesia ferruginea*).

At the lower elevations (about 300 to 1000 m asl) around Twin Sisters, the ultramafic soil parent materials are generally covered by glacial till and layers of volcanic ash, which are commonly mixed with ultramafic materials in colluvium (Goldin 1992). The till and ash are commonly thin enough, or ultramafic materials are sufficiently concentrated (but not too concentrated) in the colluvium, that ultramafic Spodosols (Cryorthods) have formed and dominate the landscape around the western margin of Twin Sisters. The natural vegetative cover on the ultramafic Spodosols is conifers and shrubs, and the timber site index for growth of western hemlock is about 17 or 18 m in 50 years (Goldin 1992).

Kruckeberg (1969) found no local endemics, but the widespread regional serpentine indicators *Aspidotis densa* and *Polystichum lemmonii* were common. Between 1500 and 1610 m asl near Orsano Creek, he described several vegetation types identified by *Pinus contorta* var *latifolia, Abies lasiocarpa, Tsuga mertensiana, Callitropsis nootkatensis, Phyllodoce glanduliflora, Silene acaulis, Sibbaldia procumbens*, *Luetkea pectinata*, and *Juniperus communis*. Other than the rather high elevation of the *Pinus contorta* var *latifolia* at timberline in the Twin Sisters area, Kruckeberg described the floristic contrasts with nonultramafic areas as less pronounced than in the Ingalls complex area.

Figure 8-10. Juniper grassland on ultramafic lithic Inceptisols in Washington Park, Fidalgo Island.

8F2 Eastern San Juan Islands

A stack of several oceanic terranes was thrust westward over the Wrangellia terrane during the Cretaceous (Brandon et al. 1988). The Decatur terrane at the top of the stack contains the Fidalgo ophiolite. On Fidalgo Island, Brown (1977) has described a complete ophiolite section that was considered to be Jurassic. The ultramafic rocks are mainly mildly serpentinized dunite and harzburgite that has a tectonite fabric. Ultramafic rocks are exposed in a linear trend from Cypress Island on the north to Fidalgo Head on Fidalgo Island and to Burrows and Allan Islands south of Fidalgo Head. The topography has been modified by regional glaciation during the Quaternary. The last glaciation in the northern Puget Sound area, the Vashon stade (stage) of the Fraser glaciation, reached a maximum extension near Seattle about 17 ka ago (Porter and Swanson 1998). Glaciation has left low, rounded hills that are generally elongated from north to south (or more precisely NNE-SSW, with local exceptions), which was the direction of ice flow. The rise of sea level (~130 m globally) from melting glacial ice would have inundated this terrain, but isostatic rebound from weight loss following melting of a 500 m cover of ice (about 450 Mg/m^2, or 50 tons per square foot) has countered the sea level rise.

The current mean annual precipitation at Anacortes on Fidalgo Island is about 660 mm, with only 24 mm in July when the mean daily temperature is 22.5°C. Small amounts of snow fall during winters, but it melts between snow-fall events.

Deep soils of the Guemes series (Haploxeralfs) are the main ones in the glaciated ultramafic terrains of the eastern San Juan Islands (Klungland and McArthur 1989). On Fidalgo Island, the ultramafic rocks are exposed on and around Fidalgo Head, which is on the northwest corner of the island, where they are limited to a narrow fringe from the shoreline inland to no more than a few hundred meters. The ultramafic soils on Fidalgo Island are mostly lithic Mollisols (Lithic Haploxerolls) with much bedrock showing through the soil cover. The vegetation on the shallow ultramafic soils is Rocky Mountain juniper (*J. scopulorum*) grassland (Fig. 8-10), with much tufted phlox (*Phlox diffusa*) and serpentine fern (*Aspidotis densa*). Denser clumps of trees occupy deeper, but still relatively shallow ultramafic soils, in and around the juniper grassland. Trees in the clumps are Douglas-fir, Rocky Mountain juniper, and madrone (*Arbutus menziesii*). Blue wildrye (*Elymus glaucus*) is common under the clumped trees. The most abundant shrub in denser Douglas-fir forest of nonerpentine terrain near Fidalgo Head in Washington Park is salal (*Gaultheria shallon*), which seems to avoid the ultramafic soils.

Kruckeberg (1969) sampled relevés in "grassy balds" on Fidalgo and Cypress Islands, one representing a deeper soil and another representing a steeper stony soil. He considered vegetation on the "balds" to be *Achillea millefolium–Aspidotis densa–Daucus pusillus* associations. The deeper soil sites had relatively large amounts of the nonnative grasses *Bromus hordaceous* and *Aira caryophyllea*, as well as native grasses such as *Festuca idahoensis, Vulpia micrstachys,* and *Koehleria macrantha*. Kruckeberg did not indicate actual soil depths; by "deeper soil" he was likely referring to deeper shallow (depth<50 cm) soil. Kruckeberg (1969) has two excellent photographs of serpentine meadows and juniper, with Douglas-fir in the background, and a similar photo on the lower slopes of Olivine Hill on Cypress Island that shows beach pine (*P. contorta*) in addition to the other species. Beach pine is absent from the Fidalgo Island ultramafics.

Ryan (1988) found 61 species of lichens on Fidalgo Island. One species had not been reported in North America previously, but none of the lichens were serpentine endemic species.

8F3 Ingalls complex

The Ingalls complex (Ic in Fig. 8-2) consists primarily of mantle tectonites (MacDonald et al. 2008). Harzburgite and dunite on the south are separated from harzburgite on the north by ultramafic mélange of the east-west oriented Navaho fault zone. The Ingalls complex is similar in age and composition to the Josephine ophiolite of the Klamath Mountains (MacDonald et al. 2008). Lange blocks of nonultramafic oceanic associates of the mantle tectonites, such as the Iron Mountain and Esmeralda Peaks units, occur in the Navaho fault zone.

Ultramafic soils and plant communities were characterized near the northwest end of the Ingalls complex exposure (Ingalls sites 1, 2, and 3) and near the drier southeast end. Although the distance is only about 20 km, the mean annual precipitation difference is from about 2000 mm at 1650 m asl on the northwest to 1000 mm below 1500 m asl on the southeast, with substantial differences related topographic relief throughout the complex. Because of dry summers, the vegetation appears more xeric than in areas north of the Cascade Mountains with comparable annual precipitation.

Three soils were described in a transect from the South Fork of Fortune Creek up to the ridge between Ingalls Peak and Esmeralda Peaks where the ridge bends westward toward Hawkins Mountain (Table 8.1). The ultramafic soils east of Hawkins Mountain are predominantly Inceptisols. Spodosols develop where sufficient volcanic ash has been deposited over the ultramafic materials to supply the Al (aluminum) for a spodic horizon. Ultramafic soils and plant communities described at Ingalls sites 1, 2, and 3 (Table 8.2) appear to be among the more common ones southwest of Ingalls Peak. Soils and vegetation at the three sites are (1) a very shallow Entisol with a sparse forb plant community at 1805 asl, (2) a moderately deep Inceptisol with an open Douglas-fir–subalpine fir–whitebark pine/grouseberry (*Vaccinium scoparium*) plant community at 1790 m asl, and (3) a lithic Inceptisol with a low juniper (*J. communis*)/bluebunch wheatgrass (*P. spicata*) plant community at 1650 m asl. Some of the plant communities on the Inceptisols have much pinemat manzanita (*A. nevadensis*) and pink mountain heath (*Phyllodoce empetriformis*) in the understories. Volcanic ash has been incorporated into the surface of the Inceptisols, resulting in moderately acid (pH 5.8) surface soils, which is unusually low for ultramafic Inceptisols in western North America. Ultramafic colluvium on lower footslopes west of Ingalls Peak, in the Fortune Creek drainage, generally have dense mixed conifer forest plant communities with much black huckleberry (*V. membranaceum*) in the understory. Conifers in the mix are Douglas-fir, hemlock (*Tsuga* sp), Engelmann spruce *(P. engelmanni*), and fir (*Abies* sp). It is difficult to ascertain which footslope soils have enough ultramafic material to be considered ultramafic where nonultramafic colluvium and glacial drift have been mixed in with the ultramafic materials.

Kruckeberg (1979) found that, in the Ingalls area, Lemmon's fern (*P. lemmonii*), serpentine fern (*A. densa*), little mountain bluegrass (*Poa curtifolia*), and several forbs were restricted to ultramafics, generally sparsely vegetated ultramafics. He named several forbs and a shrubby willow (*Salix brachycarpa*) that were good serpentine indicators. Also, he found that yarrow and several other plant species that grow in nonultramafic soils have developed strains that grow in ultramafic soils (Kruckeberg 1967). Two of the species that developed serpentine tolerant strains are nonnative weeds, self heal (*Prunella vulgaris*) and sheep sorrel (*Rumex acetosella*).

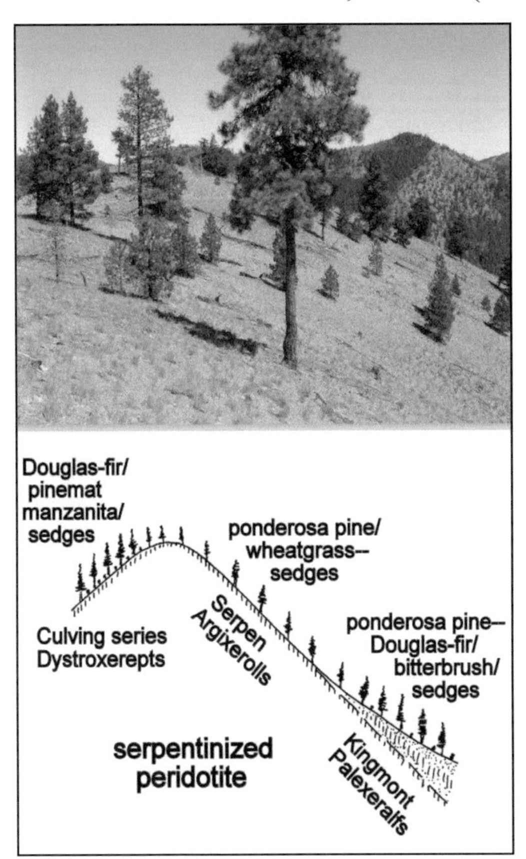

Figure 8-11. A landscape profile across serpentinized peridotite of the Ingalls complex near Peshastin Creek

At lower elevations on the east end of the Ingalls complex, ultramafic soils have been mapped by T. Aho and others of the Natural Resources Conservation Service (R. Myhrum, personal communication, 2003). Moderately deep Inceptisols (Culving series, Dystroxerepts), deep Mollisols (Serpen series, Argixerolls), and very deep Alfisols (Kingmont series, Palexeralfs) were mapped on ultramafics below about 1500 m in the Ingalls complex (Fig. 8-11). Slopes facing north to northeast are dominated by the Culving soils, with Douglas-fir (*P. menziesii*)/pinemat manzanita (*A. nevadensis*)/sedge plant communities containing some ponderosa pine (*P. ponderosa*) on drier slopes and some white pine (*P. monticola*) on cooler or more moist slopes. Lodgepole pine (*P. contorta*) is a common successional species following severe fires. Occurrences of Oregon grape (*Mahonia aquilinum*) and spiraea (*S. betulifolia*) are spotty or sparse. Slope aspects other than north to northeast are dominated by Mollisols (Argixerolls) with open to sparse conifer forest. The conifer forests are either ponderosa pine and Douglas-fir, with bitterbrush (*Purshia tridentata*) and bluebunch wheatgrass, or ponderosa pine with bluebunch wheatgrass. Sulphur buckwheat (*Eriogonum umbellatum*) and other buckwheat species (possibly *E. compositum* var *lancifolium*), balsamroot (*Balsamorhiza sagittata*), and lupine (*Lupinus* sp) are common in the open conifer forests on argillic Mollisols (Argixerolls). These argillic Mollisols are generally moderately deep to shallow on convex slopes and deep to very deep on concave slopes. Where the argillic horizons are deep in colluvium, the soils are in the Kingmont series of Alfisols.

Table 8.2 Soil texture (field grade) and laboratory data for sites characterized in Table 8.1.

Soil Horizons	Depths (cm)	Field Texture	Ca/Mg (molar)	pH
Livengood 1 (altitude 550 m) - loamy-skeletal over fragmental, magnesic Typic Haplocryept				
A, Bw, 2Bw	0-4, -16, -36	SiL, GrSiL, xGrSL	0.43, 0.24, 0.23	5.8, 6.6, 6.8 (I)
Livengood 2 (altitude 400 m) - loamy-skeletal over fragmental, magnesic Typic Haplocryept				
A, Bw, 2Bw	0-2, -22, -54	GrSiL, GrSL, xGrSCL	0.30, 0.22, 0.11	6.8, 7.0, 7.2 (I)
Shulaps 1 (altitude 2450 m) - loamy-skeletal, magnesic Typic Cryorthent				
A1, A2, C	0-7, -22, -35	xCbLS, CbL, xCbSL	0.22, 0,13, 0.08	7.2, 7.0, 7.0 (E)
Shulaps 2 (altitude 2150 m) - loamy-skeletal, magnesic, Vitrandic Haplocryept				
2Bw1, 2Bw2	62-66, -85	—	0.11, 0.09	6.2, 6.3 (E)
Shulaps 3 (altitude 1750 m) - loamy-skeletal, magnesic, Typic Haplocryept				
E, 2Bs, 2C	0-9, -24, 44-62	SL, vGrL, vGrSL	0.67, 0.30, 0.08	5.4, 5,1.6.6 (E)
Shulaps 4 (altitude 1410 m) - loamy-skeletal, magnesic, frigid Vitrandic Eutrudept				
2Bw1, 2Bw2	20-28, -40	—	0.08, 0.04	6.7, 6.8 (E)
Shulaps 5 (altitude 1370 m) - loamy-skeletal, magnesic, frigid shallow Hapludalf				
A, 2Bt, 3Bt	0-3, -18, -27	GrLS, vGrSCL, vGrCL	1.40, 0.42, 0,41	6.6, 6.8, 6.8 (E)
Tulameen 1 (altitude 1730 m) loamy-skeletal, magnesic, Typic Haplocryept				
A, Bw, C	0-15, -20, -26	—	0.18, 0.14, 0.07	6.3, 6.2, 6.6 (E)
Tulameen 2 (altitude 1480 m) - loamy-skeletal, magnesic, frigid Lithic Haplocryoll				
A1, A2. A3	0-5, -27, -44	vGrL, vGL, xGrL	0.15, 0.11, 0.12	6.8, 6.9, 7.0 (E)
Tulameen 3 (altitude 1460 m) - loamy-skeletal, magnesic, frigid Typic Haplocryept				
A1, Bw, C	0-6, -60, -75	vGrSL, vGrL, xGrSL	0.25, 0.10, 0.05	6.6, 6.8, 6.8 (E)
Tulameen 4 (altitude 1325) - loamy-skeletal, magnesic, frigid Pachic Hapludoll				
A1, Bw, 2C	0-9, -42, -99	vGrL, xStL, vGrL	0.30, 0.07, 0.07	5.8, 6.2, 6.4 (E)
Tulameen 5 (altitude 1260) - loamy-skeletal, magnesic, frigid Dystric Eutrudept				
A, Bw, C	0-17, -26, -?	—	0.34, 0.07, 0.05	6.2, 6.7, 6.8 (E)
Ingalls 1 (altitude 1805 m) - loamy-skeletal, magnesic Lithic Haplocryent				
A, Bw	0-3, -16	vStSL vGrSL	—	6.9, 7.0 (I)
Ingalls 2 (altitude 1790 m) - loamy-skeletal, magnesic Haplocryept				
A, E/B, 2Bw	0-3, -15, -30	GrfSL GrfSL, vGrL	1.53, —, 0.32	5.8, 5.9, 6.0 (I)
Ingalls 3 (altitude 1650 m) - loamy-skeletal, magnesic Lithic Haploxerept				
A, Bw, C	0-5, -22, -42	GrL, vGrL, vGrSL	—	6.8, 6.9, 7.0 (I)

Texture:: Cb, cobbly; CL, clay loam; f, fine; Gr, gravelly; L, loamy; LS, loamy sand; S, sand; SC, sandy clay; SiL, silt loam; SL, sandy loam; St, stony; v, very, x, extremely stony. pH with bromthymol blue indicator (I) or with a glass electrode (E).

9 Central Pacific Cordillera

The Sierra Nevada, Klamath, and Blue Mountains (Fig. 9-1) are major mountain ranges in western North America that have somewhat similar geologic histories. Similarities between the Klamath Mountains and the Motherlode, or "Foothill", sector of the Sierra Nevada are commonly recognized by geologists (Mankinen and Irwin 1990). The Klamath Mountains are offset westward from the Sierra Nevada, and the Blue Mountains have been rotated clockwise and are separated from the Klamath Mountains by the Cascade Mountains. The ultramafic rocks are all in oceanic terranes, with the possible exception of the Monumental Ridge body in the Sierra Nevada.

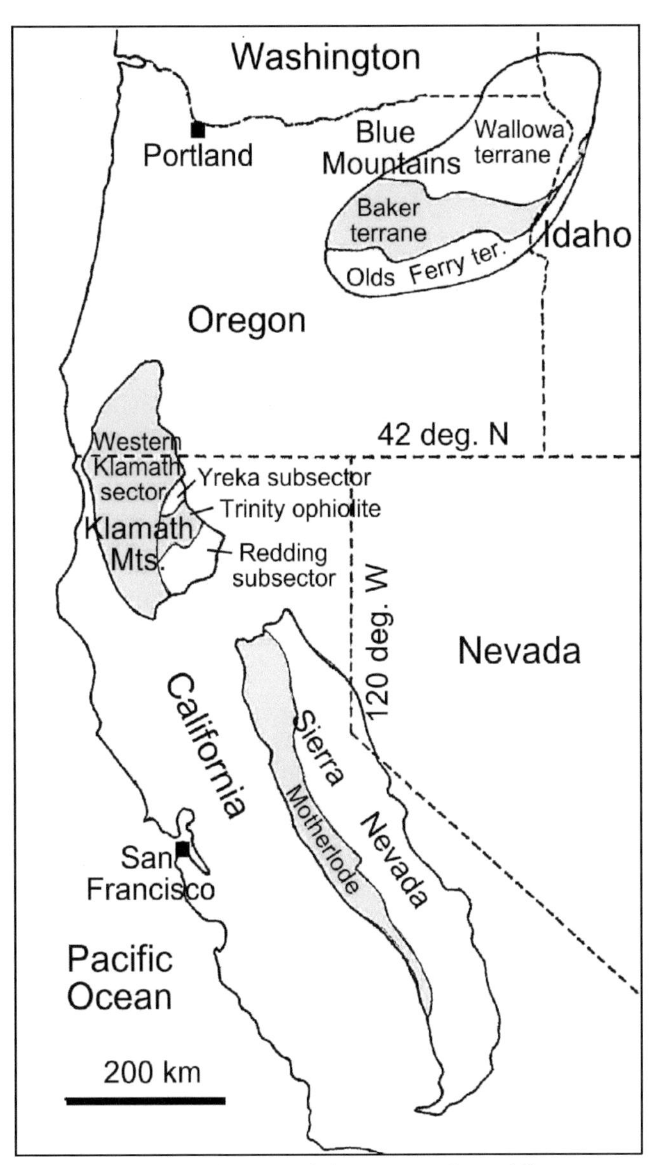

Figure 9-1. Provinces of the Central Pacific Cordillera with areas containing most of the ultramafic rocks shaded gray.

Some exotic strata of the Central Pacific Cordillera were added to a precursor of North America from about 0.4 to 0.25 Ga, but most of the allochthonus strata and the autochthonus plutons were added later, mostly in the Mesozoic era. Most of the ultramafic rocks are in steeply sloping mountain terrains, where erosion has limited soil development. The soils are mostly Alfisols and Mollisols, some Inceptisols, and few Entisols, Vertisols, and Ultisols; and there are a few Histosols over ultramafic sediments in the Klamath Mountains. Miocene surfaces of the old Klamath Peneplain that have not succumbed to erosion, however, are more than 5 million years old (Diller 1902, Aalto 2006) and they have well developed soils.

Climates range from very cold (cryic soil temperature regime, STR) in the higher mountains to cold (frigid STR) and cool (mesic STR), with substantial areas of warm (thermic STR) soils in the Motherlode sector of the Sierra Nevada.

The summers are dry (Fig. 9-2), with xeric soil moisture regimes (SMRs) and relatively small areas of udic SMRs. Substantial areas of ultramafic landscapes in the Klamath Mountains were glaciated during the Quaternary. Practically all ultramafic terrain in the Sierra is too l The summers are dry (Fig. 9-2); the soil moisture regimes (SMRs) are mostly xeric, with relatively small areas of udic SMRs. Practically all ultramafic terrain in the Sierra is too ow to have been glaciated, with the exception of a small area from the ridge west of the Red Mountain summit in Plumas County northward down to Grizzly Lake, although the till there contains both ultramafic rocks and plenty of nonultramafic metamorphic rocks. Small areas have been glaciated at the heads of stream drainages in ultramafic terrain on the north side of Canyon Mountain, in the Strawberry Range of the Blue Mountains (Brown and Thayer 1966).

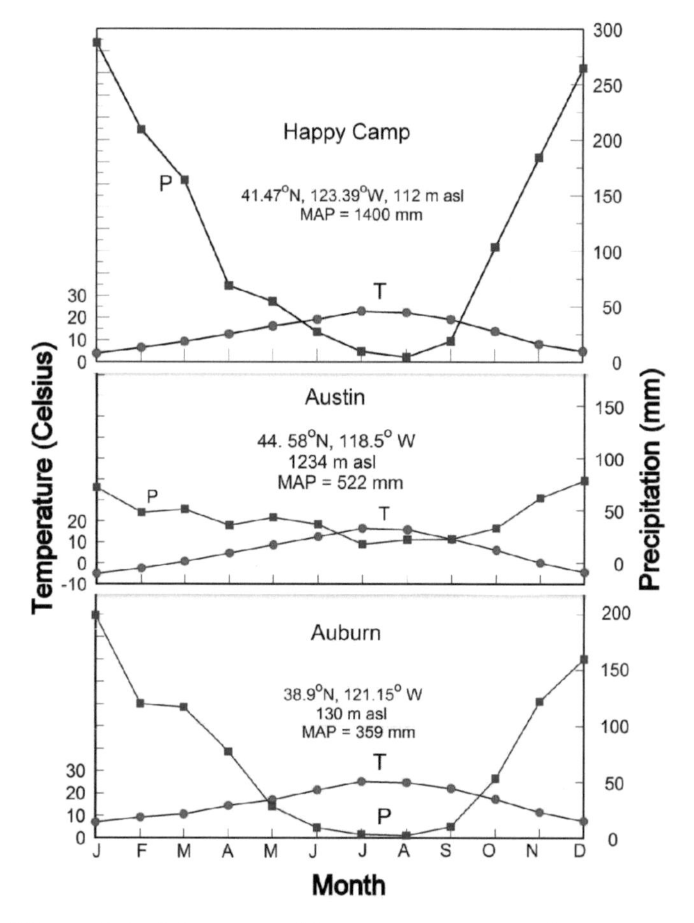

Figure 9-2. Mean monthly temperatures and precipitation at Happy Camp in the Klamath Mountains, Austin in the Blue Mountains, and Auburn in the Motherlode of the Sierra.

The diversities of ultramafic soils and serpentine plants are great in the Central Pacific Cordillera area. Plant diversity is particularly high in the Klamath Mountains. Safford et al. (2005) have listed plant species that are more common on ultramafic than on other substrates in California. Although ultramafic rocks occupy no more than 1% of the area in California, 13.4% of the endemic plant species are restricted entirely, or mostly, to ultramafic substrates (Safford 2011).

A. Geology

Major tectonic activity along the central western margin of the North American craton that began in the middle of the Paleozoic with the Antler orogeny and was followed later by the Sonoma orogen toward the end of the Paleozoic era is evident east of the Sierra Nevada, between the Sierra and the North American craton. The westward limit of the early Paleozoic craton is shown

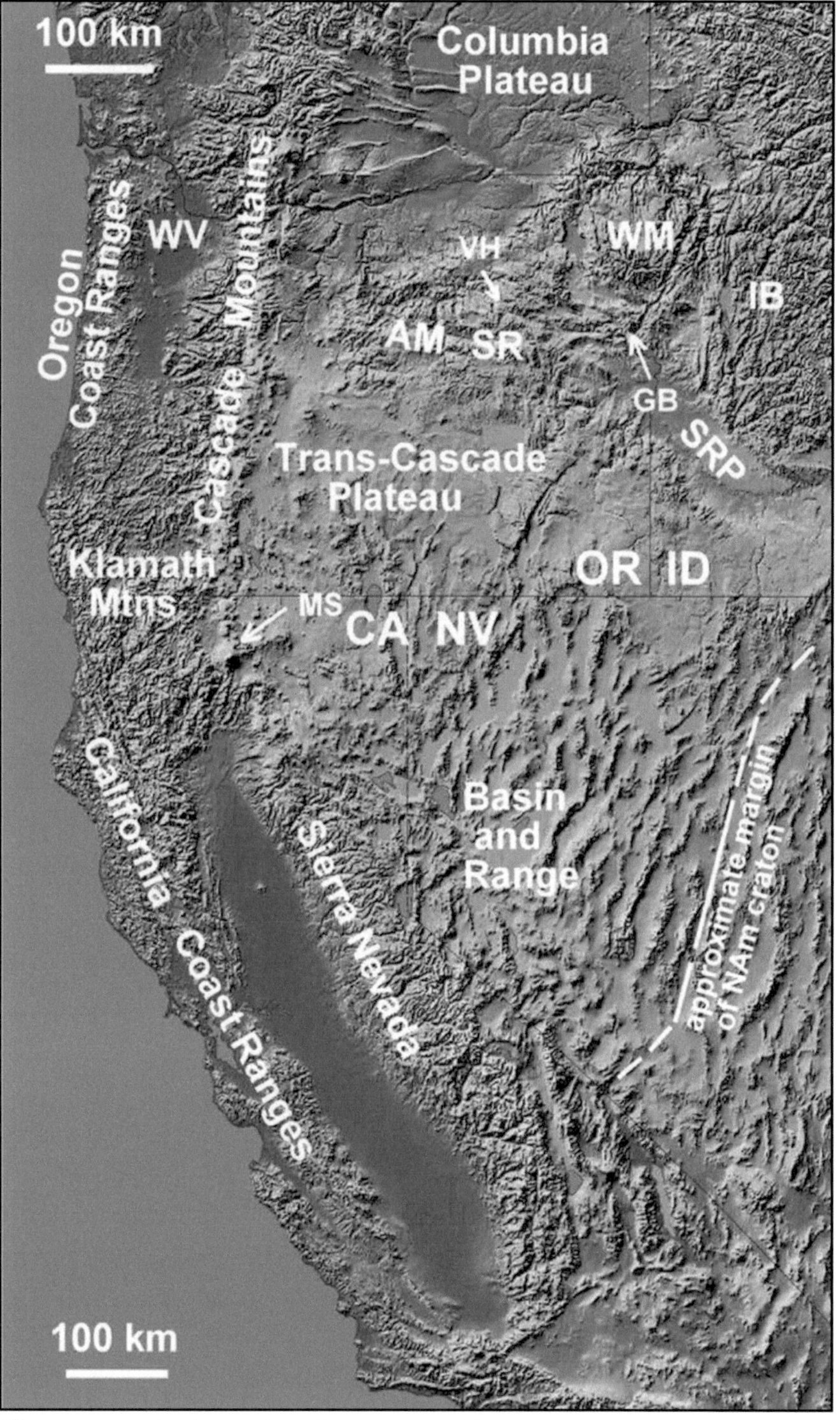

Figure 9-3. Physiography of Central Pacific Cordillera area. Notice expansion of the map scale bar with higher latitude. AM, Aldrich Mountains; GB, Glasgow Butte; IB, Idaho Batholith; MS, Mount Shasta; SR, Strawberry Range; SRP, Snake River Plain; VH, Vinegar Hill, WM, Wallowa Mountains; WV, Willamette Valley.

in Nevada (Fig. 9-3) and it extends northward (not shown) along the western side of the Idaho batholith (IB). Rocks accreted during the Antler and Sonoma orogenies are largely pericratonic, having only minor bits of oceanic crust with ultramafic rocks. The most notable exposure of ultramafics east of the Sierra Nevada is at Candelaria (Page 1959). The ultramafic landscapes at Candelaria have been drastically altered by mining activities.

Most of the ultramafic rocks in the Central Pacific Cordillera are in oceanic terranes that were accreted to the continent during the Mesozoic era. The accreted terranes in the Sierra Nevada, Klamath, and Blue Mountains are separated by major fault systems into belts, or sectors, that have distinctively different geological histories and compositions.

Sierra Nevada geology, the ultramafic rocks. The dominant feature of the Sierra Nevada is the granitic batholith (Fig. 9-4). It consists mainly of Cretaceous granitic plutons and some mafic plutonic rocks.

Prior to emplacement of the Cretaceous plutons the continental margin was occupied by pericratonic sedimentary and volcanic strata that are represented by those strata in the Shoo Fly complex east of the Melones fault and by the Cadaver and Smartville complexes west of the Melones fault (Day et al 1985). The Cadaver complex and adjacent Feather River belt consist of oceanic terranes that were accreted during the Triassic and the Smartville is a volcanic arc terrane added to the continent during the Jurassic period. Most of the ultramafic rocks in the Sierra are in the Cadaver complex and Feather River belt (Fig. 9-4). Ultramafic rocks, peridotite and serpentinite, are most concentrated in the Feather River belt, which is strip a few kilometers wide from the North Fork of the Feather River southward more than 100 km, beyond Forest Hill and across the Middle Fork of the American River. Ultramafic bodies along the Kings River, east of Fresno, and south of the Kaweah River (Fig. 9-4) might be related to those in the Cadaver complex (Saleeby 2011).

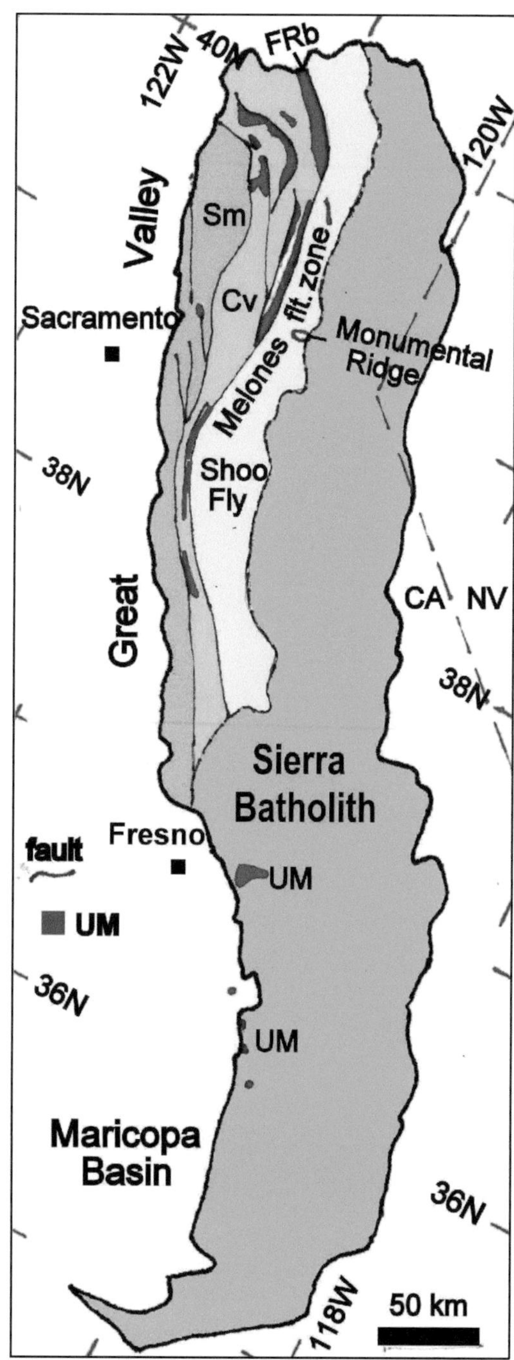

Figure 9-4. Figure 9-4. The Sierra Nevada following a map of the California Division of Mines and Geology (1966). Cv, Cadaver complex; FRB, Feather River belt; Sm, Smartville complex; UM, isolated ultramafic bodies (dark red).

Late Cretaceous and Cenozoic erosion has removed much stratigraphic cover from the Sierra batholith and exposed granitic plutons both east and west of the Melones fault zone. The contrast in physiographic development of the northern and the southern Sierra is great, however, and the southern Sierra lacks the extensive pyroclastic and volcanic mudflow (lahar) deposits that are common across the northern Sierra (Wakabayashi and Sawyer 2001).

Klamath Mountains geology, the ultramafic terranes. Ultramafic rerranes are widely distributed across the Klamath Mountains (Fig 9-5). The Eastern Klamath is comprised of late Proterozoic, Paleozoic, and Mesozoic terranes that were accreted to North America during the Mesozoic era. Ultramafic rocks of the Eastern Klamath are in the Trinity ophiolite, with ultramafic rocks absent from the Yreka subsector, which is between Yreka and the Trinity ophiolite, and absent from the Redding subsector southeast of the Trinity ophiolite. Ultramafic rocks are common along the Trinity fault between the Eastern and Western Klamath, as far north as Yreka (Fig. 9-5), but absent along the Bully Choop fault to the south. There are many large and small ultramafic bodies in the Western Klamath, including the large Josephine ophiolite along the western edge of the Klamath Mountains. Some rocks, including ultramafic rocks, have been thrust from the Klamath Mountains across the California Coast Ranges.

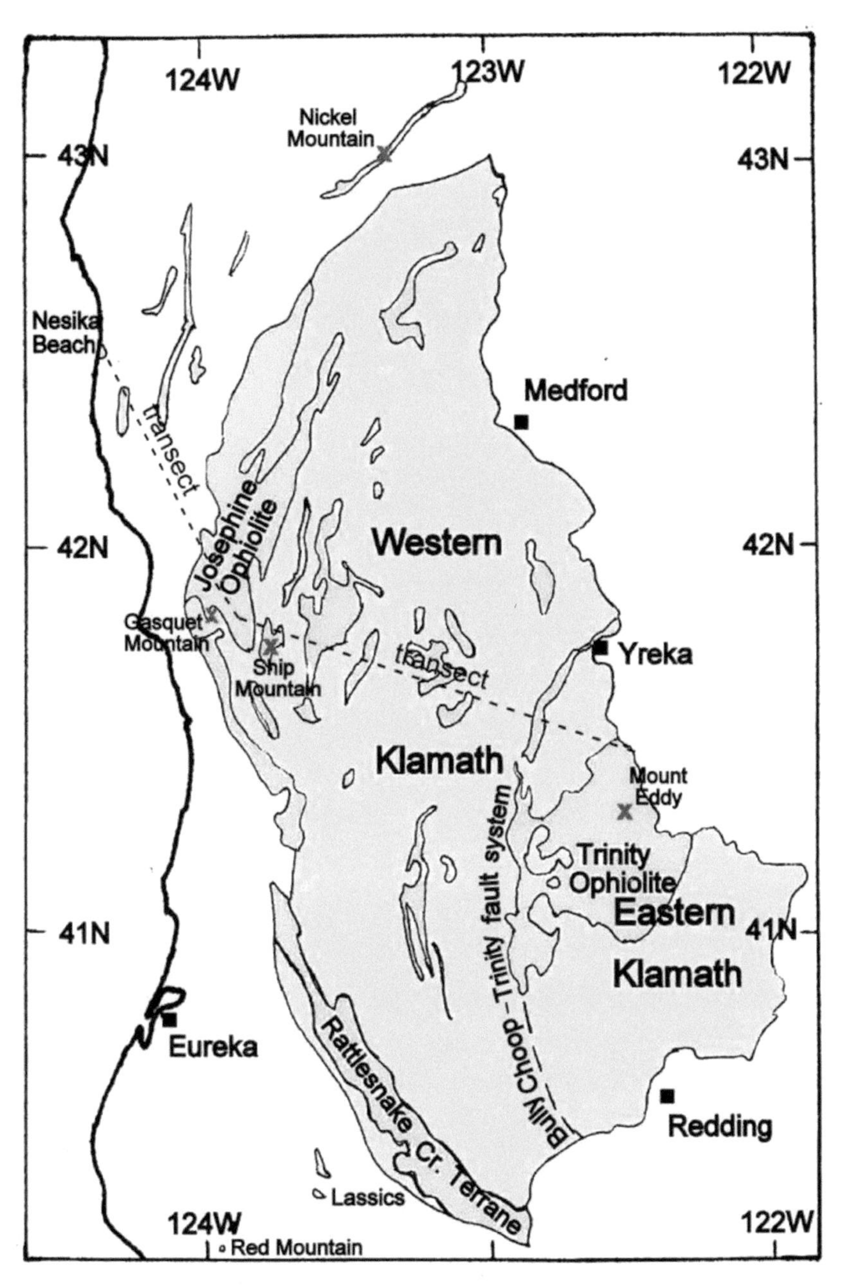

Figure 9-5. Ophiolitic terranes of the Klamath Mountains, based on maps of Irwin (1981), Jennings (!977), and Walker and MacLeod (1991). Rough and Ready Creek is on the eastern edge of the Josephine ophiolite.

Blue Mountains geology. A comprehensive treatment of the Blue Mountains geology was published in the latter part of the 20th century (Vallier and Brooks 1995). Subsequently, there has been substantial progress in developing models for the tectonic history (Dorsey and LaMaskin 2007, Schwartz et al. 2010). Although there are several scenarios that differ in details, they all involve two magmatic arc terranes (Wallowa and Olds Ferry), and an oceanic subduction and accretionary complex (Baker terrane) between them (Dorsey and LaMaskin 2007). Ultramafic rocks of the Blue Mountains are concentrated in the Baker terrane (Fig. 9-1). During the Triassic and Jurassic periods, thrust faulting between the converging Olds Ferry and Wallowa terranes modified the already complex Baker terrane. These terranes were attached to the North American continent late in the Jurassic and subsequent plutonic intrusions have amalgamated them.

Oceanic terranes of the Blue Mountains were covered by Cenozoic volcanic flows, pyroclastic deposits, and sediments. Subsequently, late Cenozoic uplift and geologic erosion have exposed large areas of ultramafic rocks. The largest areas of exposed ultramafic rocks are in the Aldrich Mountains and the Strawberry Range, just south of the John Day River, and in many smaller areas around Vinegar Hill.

B. Sierra Nevada

Altitudes of ultramafic exposures range from about 100 m asl (above sea level) adjacent to the Great Valley to 1935 m on Red Hill and 1939 m on Red Mountain, which are in the Plumas National Forest. The precipitation ranges from about 750 to more than 2000 mm/year. Summers are dry (Fig. 9-2, Auburn). The ultramafic soils are mostly warm (thermic STRs) and cool (mesic STRs), with small areas of cold soils (frigid STRs) on Monumental Ridge and on Red Mountain, and the soils all have xeric moisture regimes.

The vegetation on ultramafics in the Sierra is less different from vegetation on nonultramafic substrates than are those differences in the Klamath Mountains or in the California Coast Ranges; and there is a larger proportion of exotic (non-native) forbs (Kruckeberg 1984), and there are fewer serpentine endemic species in the Sierra (Stebbins 1984). Gabbro on Pine Hill, about 40 km north-northeast of Sacramento, however, is noted for the distinctive and endemic species there (Hunter and Horenstein 1992). Medeiros et al. (2015) have summarized information on gabbro habitats in the California floristic area.

There is a nickel hyperaccumulator (leaf Ni >1000 μg/g), milkwort jewel flower (*Streptanthus polygaloides*), that is common on warm ultramafic soils in the Sierra. It is one of only two plant species in western North America that are confirmed Ni-hyperaccumulators (Reeves et al.1983). The other Ni-hyperaccumulator species is alpine pennycress (*Thlaspi montanum* or *Noccaea fendleri*), which has three subspecies that are common on ultramafics in the Klamath Mountains and in the Blue Mountains (Chapter 4).

9B1 Shoo Fly complex and Monumental Ridge

East of the Melones fault zone most of accreted terranes are pericratonic with no major areas of ophiolitic terranes. The only relatively large exposure of ultramafic rocks is on Monumental Ridge (Fig. 9-4). It has characteristics of an Alaskan-type concentric body, but geologists are reluctant to call it such a body. A dunite core is present at the west end of the ridge, above Onion Valley, and wherlite and pyroxenite constitute the backbone of Monumental Ridge (James 1971). The ultramafic rocks of this concentric body lack serpentine.

Much of the wehrlite and clinopyoxenite on Monumental Ridge is exposed on the south slope from the summit at 2092 m asl down to about 1650 meters. The high specific gravity of wherlite near the summit (3.3 Mg/m^3) verifies the lack of serpentinization.

Bedrock exposures dominate the summit of Monumental Ridge, with dense huckleberry oak (*Q. vaccinifolia*)–silktassel (*Garrya fremontii*)–greenleaf manzanita (*A. patula*) chaparral between exposures of bedrock. A few small patches of shallow soils along the ridge have a cover of pinemat manzanita (*A. nevadensis*) and prostrate ceanothus (*C. prostratus*). Talus that is barren, except for abundant crustose lichens, extends downward from just below the summit, with some dense patches of huckleberry oak–silktassel chaparral that grade on downward to huckleberry oak–Sierra coffeeberry (*Frangula rubra*) chaparral. Soils below the talus are steeply sloping, cold Inceptisols with semidense mixed conifer forest and an understory of huckleberry oak and chokecherry (*Prunus virginiana*). Trees in the mixed conifer forest are Jeffrey pine, incense cedar, sugar pine, Douglas-fir, and white fir. Stony, shallow (lithic) soils on broad spur ridges have open Jeffrey pine–incense cedar forests with huckleberry oak, pinemat manzanita, prostrate ceanothus, and sparse Idaho fescue (*F. idahoensis*) in the understory. The ceanothus in the lower plant communities has leaves that are more characteristic of prostrate ceanothus, than the smaller leaves of the prostrate ceanothus along the summit of Monumental Ridge, which might be a different species.

The largest exposure of dunite in this locality is just east of Onion Valley at about 1500 to 1650 m asl. Cool Alfisols (Mollic and Lithic Mollic Haploxeralfs) have been mapped in this area (Hanes 1994). Most of the dunite exposure is on a steep, rocky, west-facing slope. It is covered by a Jeffrey pine, incense cedar, Douglas-fir, bay (*Umbellularia californica*) forest with deerbrush (*Ceanothus integerrimus*) in the understory. Deerbrush generally avoids ultramafic soils elsewhere. There are no oak trees, but sparse huckleberry oak (*Q. vaccinifolia*) occurs on colluvial footslopes. Some of the forbs were mentioned by Alexander et al. (2007a).

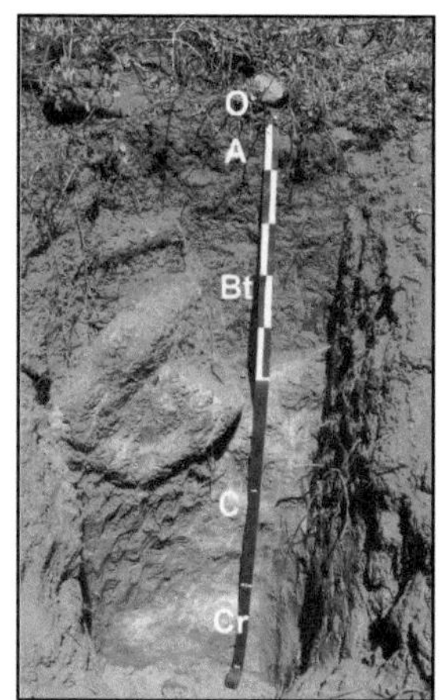

Figure 9-6. A deep soil described and sampled on Plumas Red Hill, with data in Tables 9.1 (RH1) and 9.2.

9B2 Feather River belt

Peridotite is the dominant ultramafic rock in the Feather River belt, and there is substantial dunite. Soils near the summit of the Plumas Red Hill (1935 m asl, one of many Red Hills), which is between the North and Middle Forks of the Feather River, are likely cold, but they are scarcely any different from the cool soils at lower elevations. Moderately deep Alfisols with open Jeffrey pine–incense cedar–Douglas-fir/buckbrush/ Idaho fescue stands are common on steep slopes of the Feather River belt and there are some deep Alfisols, and possibly Ultisols (Churchill 1988), on more gentle slopes such as some of those on Red Hill (Fig. 9-6).

Table 9.1 Plumas Red Hill soils; subsoil colors and laboratory data.

Pedon and Horizon	Depth (cm)	H:L:M proportions	Munsell color (m)	Fe	Ni	K	Ca/Mg (g/g)	pH DW
				g/kg (in aqua regia digest)				
RH2, moderately deep Alfisol (Haploxeralf)/serpentinite								
RH2-4, Bt1	21-33	18:76:06	10YR 4/3	58.0	2.4	0.5	0.07	5.6
RH2-5, Bt2	33-54	07:87:06	10YR 4/3	80.3	3.3	0.5	0.10	5.7
RH3, moderately deep Alfisol (Haploxeralf)/peridotite, partially serpentinized								
RH3-4, Bt1	22-35	45:50:05	2.5YR 4/5	114.7	3.1	0.8	0.02	5.6
RH3-5, Bt2	35-60	35:59:06	2.5YR 4/5	113.1	3.1	0.9	0.02	5.6
RH1, deep Alfisol (Ultic Haploxeralf)/peridotite, partially serpentinized								
RH1-5, BC	70-135	36:61:03	7.5YR 5/8	174.4	4.1	0.8	0.03	5.5

H:L:M proportions are the relative weights of heavy nonmagnetic (H), light (L), and heavy magnetic (M) grains in the fine sand (0.125-0.25 mm) separates by bromoform (SG=2.89) flotation and a hand magnet. The chemical element analyses (Fe, Ni, K, Ca, and Mg) are from aqua regia digestion.

Most of the ultramafic rocks on Red Hill are peridotite, with some relatively small areas of serpentinite. A comparison of two moderately deep soils (Table 9.1) with open canopies of Jeffrey pine, incense cedar, Douglas-fir, and sugar pine, and with buckbrush (*C. cuneatus*) on both soils, shows that the peridotite soil (RH3) is redder than the serpentinite soil (RH2). The redness reflects the amount of secondary Fe-oxides, which is related to the heavy nonmagnetic mineral content (H in the third column of Table 9.1). The heavy nonmagnetic mineral fraction contains olivine and pyroxenes, which have more Fe than minerals in the light fraction (L) and are more readily weathered than serpentine, the main mineral in the light fraction. The deep soil on Red Hill (Table 9-1) is red (2.5YR hue) in the upper B horizon, but less red (7.5YR hue) in the BC horizon where less of the Fe is oxidized from goethite to hematite, which is redder.

Table 9.2 Plumas Red Hill, a deep Alfisol (pedon RH1), soil laboratory data.

Soil Horizons		Exchange. Cat.			Extr. Acid.		Base Satn.	Ca/Mg ratio	Org. Mat. (LOI)	pH BTB
Sym.	Depth	Ca	Mg	K	pH7	pH8				
	cm	$mmol_+$/kg					%	molar	g/kg	
A1	0-9	46	48	3.5	20	106	47	0.96	58	6.2
A2	9-22	18	42	3.1	17	78	43	0.43	34	6.2
Bt1	22-36	7	40	–	14	66	42	0.18	20	6.2
Bt2	36-70	6	57	–	14	75	46	0.11	–	6.3
BC	70-135	5	80	–	14	85	50	0.06	–	6.4
C	135-152	5	101	–	17	99	52	0.05	–	6.4

Na-dithionite reduced Fe and Mn extracted in Na-citrate solution from the Bt1 horizon was 98 g Fe and 1.2 g Mn per kg of soil. Soil pH by bromthymol blue indicator (BTB) is much higher for the BC horizon of the deep Plumas Red Hill soil than it is with a glass electrode in distilled water (Table 9.1, RH1-5).

A very old ultramafic soil on the Forest Hill divide, near the south end of the Feather River belt, is on serpentinized peridotite that was covered by lahar about 4 or 5 Ma. The lahar was removed by erosion and now the soil is an ultramafic Ultisol that has a kandic horizon (Alexander 2010). It has been mapped as the Forbes series (Hanes 1994). These soils are timber producers with dense mixed conifer forests (Fig. 3-5B). Residue from the Neogene lahar cover has had positive effects on fertility of the Forbes soil. Forests on the Forbes soil have a Douglas-fir, ponderosa and sugar pine, incense cedar and white fir overstory, with black oak trees in the understory. Black oak (*Q. kelloggii*) is absent from younger ultramafic soils.

9B3 Cool and cold Motherlode landscapes

Cool ultramafic soils in the Motherlode country are practically all north from the American River and the cold soils are on Red Mountain in Plumas County. The predominant cool soils are steeply sloping, moderately deep Alfisols, and the plant communities are commonly open-canopy forests of Jeffrey pine and incense cedar with buckbrush and Idaho fescue in the understories. Steeply sloping lithic Alfisols have sparse trees and shrubs. Soils on the very steep slopes are mainly Entisols and lithic Inceptisols and Alfisols with shrubs, or chaparral, and few trees. Among the more common shrubs are buckbrush (*Ceanothus cuneatus*) and whiteleaf manzanita. Leather oak (*Q. durata*) and McNab cypress (*Hesperocyparis macnabiana*), which are serpentine endemic or mostly endemic shrubs, occur on some of the soils.

The cold ultramafic soils are on Red Mountain in Plumas County. They are on a ridge four kilometers long, with altitudes about 1850 to 1900 m along the ridge. The cold ultramafic soils on the ridge are mostly lithic and moderately deep Inceptisols. Dense stands of red fir with sparse to common western white pine occupy most of the deeper soils, with pinemat manzanita (*A. nevadensis*) where the forest is more open. Rocky and very stony areas are occupied by Jeffrey pine and huckleberry oak (*Q. vaccinifolia*), with yellow rabbitbrush (*Chrysothamnus viscidiflorus*) in the more extremely rocky areas. Between dense forest dominated by red fir and open to sparse forest with Jeffrey pine, there are semidense forests with plenty of western white pine. Greenleaf manzanita (*A. patula*) is common in the understories; and coffeeberry (*Frangula rubra*), serviceberry (*Amelanchier alnifolia*), spiraea (*S. douglasii*), and Scouler willow (*S. scouleriana*) are sparse. A drastically disturbed area with top soil scraped off into piles has a sparse cover of western white pine and lodgepole pine trees, pinemat manzanita, sulphur buckwheat (*Eriogonum umbellatum*), phlox (*P. diffusa*), and squirreltail (*Elymus elymoides*). The 1939 m summit of Red Mountain is on a pile of nonultramafic rocks, presumably metadiorite (Hietanen 1973), that rise three or four meters above a very gently sloping surface with a dense stand of red fir trees.

Figure 9-7. Lithic, warm (thermic STR) Alfisols in a gray pine/buckbrush/grass plant community of the Red Hills in Tuolumne County.

9B4 Warm Motherlode landscapes

Warm ultramafic soils in the Sierra foothills are mostly on steep to very steep slopes near the Feather River and on hills with more gentle slopes further south. Lithic Alfisols are common, and there are some moderately deep Alfisols. Gray pine/buckbrush/annual grass plant communities are common (Fig. 9-7), with soft chess (*Bromus hordeaceus*), wild oat (*Avena* sp.), and annual fescue (*Vulpia microstachys*) grasses. Besides the more widely distributed buckbrush and whiteleaf manzanita (*A. viscida*), toyon (*Heteromeles arbutifolia*, and shrubby bay (*Umbellularia californica*) are common on Inceptisols and Entisols of very steep north-facing slopes. Valley silktassel (*Garrya congdonii*) and chaparral pea (*Pickeringia montana*) are other plants that are widespread on the ultramafic soils. Toward the south, Mariposa manzanita (*A. viscida* ssp. *mariposa*) replaces whiteleaf manzanita, there is blue oak (*Q. douglasii*) on some north-facing slopes, and grassland lacking trees and shrubs is common.

9B5 Pine Hill gabbro

The gabbro soils on Pine Hill, northeast of Sacramento, are mostly moderately deep, warm Alfisols with chamise–manzanita and mixed chaparral, and a small area of moderately deep, cool Alfisols with black oak (*Q. kelloggii*)/toyon (*Heteromeles arbutifolia*)–Lemmon ceanothus (*C. lemmonii*) on north-facing slopes (Alexander 2011). There are four plants that are endemic on the warm gabbro soils of Pine Hill: *Ceanothus roderickii*, *Fremontodendron decumbens*, *Galium californicim* ssp *sierrae*, and *Wyethia reticulata* (Wilson et al. 2010).

C. Klamath Mountains

Altitudes of ultramafic rocks and landscapes range from about 200 m asl along the Smith River near Gasquet up to 2751 meters on Mt. Eddy of the Trinity ophiolite in the Eastern Klamath sector (Fig. 9-5). The ultramafic soils range from warm to mostly cool, some cold and few very cold soils (thermic, mesic, frigid, and cryic STRs). The precipitation ranges from about 400 to more than 3000 mm/year. Summers are dry (Fig. 9-2, Happy Camp), but soils receiving high precipitation and those exposed to fog along the Pacific coast may be humid (udic SMR).

Along with greater ultramafic habitat diversity in the Klamath Mountains, the ultramafic plant diversity is also greater than in the Sierra Nevada (Harrison 2013). The great botanical diversity was recognized by Whittaker (1960), who compared plant distributions on silicic and mafic plutonic rocks and on ultramafic rocks in the west-central Klamath Mountains.. Vegetation responses to climatic changes through the Holocene were evaluated by Briles et al. (2011), and Damschen et al. (2010) revisited Whittaker's plots to record current trends.

9C1 Rattlesnake Creek terrane

The southern exposure of the Rattlesnake Creek terrane (RCT) near the southwestern margin of the Klamath Mountains has been an excellent laboratory for the study of ultramafic soils and the soil-vegetation relationships (Alexander 2003). Prior to the separation of peridotite and serpentinite soils in the RCT, they had not been differentiated in any soil surveys of North America.. The RCT soils were mapped in detail, which is unusual for ultramafic soils, because they have negligible agricultural potential. A detailed soil survey was done in the RCT, because plant ecologists of

Figure 9-8. A Jeffrey pine-incense cedar/ buckbrush/Idaho fescue plant community on Hyampom soil, a cool Mollic Haploxeralf.

the Shasta-Trinity National Forest wanted to know the extent and characteristics of some rare and endangered plant species habitats (Julie Nelson, personal communication, 1999).

About 12,000 ha of ultramafic soils were mapped in detail in the RCT, and 196 pedons were described to characterize the ultramafic soils. They are mostly cool and warm Alfisols and Mollisols (Fig. 9-8, Table 9.3), less common Entisols and Inceptisols, sparse Vertisols, and a unique Histosol. The vegetation is mostly open conifer forest and shrubland with various shrub species. Mean annual precipitation ranges from about 500 mm in warm chaparral to 2000 mm at the higher elevations, but is mostly in the 750 to 1500 mm range.

Table 9.3 Dominant cool ultramafic soils in four Rattlesnake Creek terrane soil habitat classes.

Habitat Class	Peridotite Soils	Serpentinite Soils
very shallow soils	Entisols	haplic Mollisols
shallow soils	Alfisols	argillic Mollisols
deeper S-facing soils	Alfisols	argillic Mollisols >Alfisols
N-facing slopes	Alfisols	Alfisols > argillic Mollisols

Habitat class definitions: very shallow soils, depth <18 cm for soils with argillic horizons, otherwise depth <25 cm; shallow soils, other than very shallow soils, and depth <50 cm on S-facing slopes or depth <30 cm on N-facing slopes; deeper S-facing soils, depth >50 cm on S-facing slopes; N-facing slopes, soils >30 cm deep on N-facing slopes (azimuth 300 to 120 degrees).

The most profound differences in the ultramafic soils are that the peridotite soils have much more secondary Fe-oxides and are redder than the serpentinite soils (Alexander 2004), because most of the iron in peridotite is in olivine and pyroxenes, which are readily weatherable, and most of the iron in serpentinite is in serpentine, which is less readily weatherable, and in magnetite, which is very resistant to weathering. Representatives of the most common cool, shallow ultramafic soils in the RCT (Table 9.4) indicate how the peridotite soil with more organic matter than the serpentinite Mollisol (Bramlet series) is an Alfisol (Wildmad series), because the extra Fe-oxides in it raise the chroma in the Bt horizon (12-33 cm depth) to 4, which prevents the peridotite soil from being classified as a Mollisol (Soil Survey Staff 1999).

Physically, peridotite and serpentinite soils are very similar. Peridotite soils tend to be stonier than serpentinite soils (Alexander and DuShey 2011). Peridotite slopes are generally steeper than serpentinite slopes (Fig. 3-2) and there are more large serpentinite than large peridotite landslides that extend beyond mountain footslopes. Steep slopes on the more massive peridotite are more stable than those on tectonically sheared serpentinite, which is much more common than massive serpentinite. Soil depth distributions are about the same for peridotite and serpentinite soils (Fig.

3-3), and they are also similar to the depths of other soils of the Shasta-Trinity National forest with parent materials of metamorphic rocks (Lanspa 1993). The common moderately deep peridotite Alfisol in the RCT is Dubakela, and the more common moderately deep serpentinite Alfisol, more common than Dubakela, is Hyampom (Figures 3-1 and 9-8) , although the Hyampom soil has not been endorsed by the National Resources Conservation Service (NRCS).

Table 9.4 A comparison of the predominant shallow peridotite and serpentinite soils with mesic soil temperature regimes in the Rattlesnake Creek terrane.

Hor.	Depth	Munsell Color	[a]Field Grade (text.)	[b]OM	Exch. Bases		Exch. Acidity		pH DW
	cm				Ca	Mg	pH 7	pH 8	
		moist		g/kg	mmol+/kg				
Wildmad series, a Lithic Mollic Haploxeralf/peridotite, pedon TS42									
Oi	3-0	slightly matted pine needles							
A	0-5	7.5YR 2/2	GrSL	95	61	166	44	105	6.3
AB	5-12	7.5YR 3/3	GrSL	42	35	175	18	77	6.4
Bt	12-33	5YR 3/4	vGrL	29	22	161	15	66	6.6
R	33+	moderately fractured partially serpentinized peridotite							
Bramlet series/serpentinite, a Lithic Ultic Argixeroll, pedon TS26									
Oi	3-0	slightly matted pine needles							
A1	0-2	10YR 2/1	vGrSL	32	16	83	8	38	6.2
A2	2-12	10YR 2/1	vGrSL	23	11	82	2	29	6.4
Bt	12-26	10YR 2/2	vGrSL	18	11	96	3	29	6.5
R	26+	highly fractured serpentinite							

[a] Field grade symbols: GrSL, gravelly sandy loam; vGL, very gravelly loam; vGrSL, very gravelly sandy loam. [b] Organic matter estimated from loss-on-ignition at 360 C, with an adjustment for CD extractable iron.

Seven plant communities, among the many observed, are representative of the more common serpentine plant distributions in the RCT (Fig. 9-9). Gray pine is the common conifer on warm soils, but it generally covers <15% of any area. Chamise is a dominant shrub on the warmest ultramafic soils, which are in foothills adjacent to the Sacramento Valley. Leather oak, a mostly serpentine endemic plant, is a dominant shrub on other warm soils east of 123.2° latitude, but there is no leather oak west of about 123.2° latitude in the RCT. Buckbrush is the most ubiquitous shrub and it is the dominant one on warm soils west of 123.2° latitude. Jeffrey pine and incense cedar are the common conifers on cool soils, also with Douglas-fir on the cooler soils. There are less shrubs and is more grass on the cool soils than on the warm soils. The dominant grasses, from warm to cooler soils, are small fescue (*Vulpia microstachys*), Idaho fescue, and California fescue. Huckleberry oak is the predominant shrub on cool soils that are marginal to cold soils. For cool soils east of 123.2° latitude, there is more tree cover on peridotite soils, at least on S-facing slopes, and more grass cover on serpentinite than on peridotite soils; both peridotite and serpentinite soils were found to have about 20% shrub cover on very shallow soils to 30% on deep S-facing soils and about 10% on N-facing slopes. Tree cover on the cool N-facing slopes averaged about 35 to 40%. Grass cover averaged no more than 10% in any habitat class on either peridotite or serpentinite soils. Three of the more ubiquitous but sparse serpentine forbs are *Phacelia corymbosa* (serpentine phacelia), *Streptanthus barbatus* (jewelflower), and *Eriogonum libertini* (buckwheat). The phacelia shows no peridotite or serpentinite soil preference. Jewel-flower (*Streptanthus* spp.) was present on only 9% of the peridotite sites, but on 36% of the serpentinite sites. These species of jewelflower are not Ni-hyperaccumulators. Prostrate milkweed (*Asclepias solanoana*) was on some of the more barren south-facing slopes.

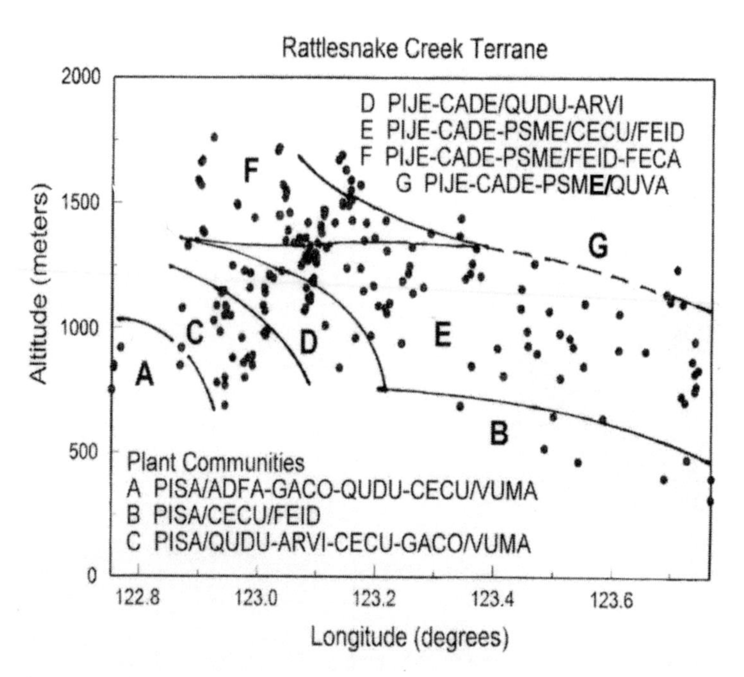

Figure 9-9. Plant communities that are representative of habitats on the warm and cool ultramafic soils in the RCT. The dots represent sites of ultramafic soil and serpentine plant community characterization. The plant species symbols are ADFA, chamise (*Adenostoma fasciculata*); ARVI, whiteleaf manzanita (*A. viscida*); CADE, incense cedar; CECU, buckbrush (*Ceanothus cuneatus*); FECA, *Festuca californica*; FEID, *Festuca idahoensis*; GACO, chaparral silktassel (*Garrya congdonii*); PIJE, Jeffrey pine; PISA, gray pine; PSME, Douglas-fir; QUDU, leather oak (*Q. durata*); QUVA, huckleberry oak (*Q. vaccinifolia*); and VUMA, small fescue (*Vulpia microstachys*).

9C2 A cool landscapes transect

Cool, moderately deep ultramafic soils (altitude <1250 m, <4100 feet) were sampled on a transect from a dry site on the edge of Shasta Valley to a humid site on the Pacific coast at Nesika Beach (Fig. 9-5). The soils were characterized to learn the effects of precipitation and soil moisture on ultramafic soils of the Klamath Mountains (Alexander 2014a). The main effects were more weathering and leaching coastward with greater precipitation and less evapotranspiration. Weathering and leaching are such dominant processes in the wetter soils that they minimize differences between peridotite and serpentinite soils. The soils, from inland to coastal, were argillic Mollisols (Argixerolls), summer dry Alfisols (Haploxeralfs), moist Alfisols (Hapludalfs), and a moist argillic Mollisol (Argiudoll). A deep, highly weathered Alfisol (Kanhapludalf, site 20) was included in the transect (Table 9.5). Most of the soils were in clayey-skeletal families.

Figure 9-10. Four sites on the arid interior to moist coastal transect across the Klamath Mountains. A. Site 01, sparse juniper/sagebrush/bluegrass on a Lithic Argixeroll, B. Site 06, Jeffrey pine–incense cedar/buckbrush/Idaho fescue on a moderately deep Argixeroll. C. Site 10, Jeffrey pine–Douglas-fir– incense cedar/Oregon white oak/Idaho fescue on a moderately deep Mollic Haploxeralf. D. Site 16, Douglas-fir–sugar pine–incense cedar/evergreen huckleberry–hoary manzanita–Pacific rhododendron–red huckleberry on a moderately deep Mollic Hapludalf.

Weathering of peridotite and serpentinite releases Si and Mg that are then leached from the wetter soils, concentrating Fe and other first transition elements from V to Ni, and also Cu and Zn, in the soils (Alexander 2014a). Smectite, the dominant clay mineral in the drier soils, is lost from the more weathered wetter ultramafic soils and the Fe that is concentrated in them forms goethite and possibly some hematite. Pea-size nodules common in the surface horizons of very old ultramafic soils are practically all magnetic, indicating the formation of magnetite or maghemite. The amounts of exchangeable cations (Mg, Ca, and K) are greatly reduced in the more leached soils. The CEC (cation-exchange capacity) decreases as smectite is lost and the AEC (anion-exchange capacity) increases as goethite becomes a more dominant clay mineral. Without measuring the AEC and CEC of a soil, a net positive charge (AEC>CEC) is indicated by higher pH in a salt solution, such as molar KCL, than in distilled water. Old ultramafic soils have net positive charges as indicated by higher pH values in a neutral salt solution than in distilled water (Alexander 2010, 2014a).

There is a large variety of ultramafic plant communities from dry inland to moist coastal habitats (Table 9.5). The plant communities range from sagebrush with sparse western juniper trees and bluegrass to dense coastal forest, with open-canopy forests at all locations between the dry and moist extremes. Western coffeeberry (*Frangula californica* ssp. *occidentalis*) which is common at many of the transect sites is endemic on ultramafic and gabbro soils.

Table 9.5 Soils and plant communities (PCs) representative of a cool soil transect across the Klamath Mountains below 1250 m asl, and with 400 to 3200 mm of mean annual precipitation.

PC	MAP (mm)	Soil (ultramafic PM)	Plant Community representing a habitat zone
A	400	Lithic Argixeroll	JUOC/ARTR/POSE, *Phlox diffusa*
B	600	Typic Argixeroll	PIJE-CADE/CECU/AGSP, *Koeleria macrantha*
C	800	Xerolls + Xeralfs	PIJE-CADE/CECU/FEID-BRLA-MEIM
D	1100	Mollic Haploxeralf	PIJE-PSME-CADE/QUGAB-CECU/FEID-BRLA
E	1600	Ultic Haploxeralf	PIJE-PSME-CADE/ARME/QUVA, *Ceanothus prostratus*
F	2300	Mollic Hapludalf	PSME-PILA/LIDE/VAOV-ARCA-RHMA/FECA
G	2800	Typic Hapludalf	PSME-PIMO/LIDE/VAOV-RHMA-VAPA-ARCA
H	3200	Typic Hapludalf	PSME-PIMO/LIDEE-UMCA-QUVA–RHMA/XETE
fire	3000	Typic Kanhapludalf	PIMO-PICO/LIDEE-ARCO-VAOV-FRCA-VAPA/XETE
I	2000	Pachic Argiudoll	PSME-ABGR/LIDE-UMCA/VAOV, Tsuga heterophylla

PC sites G, H, and “fire” might be in similar habitat zones, except that site H is higher than the others and site “fire” is on an older soil dominated by close-cone pines (*P. attenuata* and *P. contorta*) that are indicative of severe fire. The plant species symbols are ABGR, grand fir; AGSP, *Pseudoregnaria spicata*; ARCA, *Arctostaphylos canescens*; ARCO, *Arctostaphylos columbiana*; ARME, madrone; ARTR, *Artemisia tridentata*; BRLA, *Bromus laevipes*; CADE, incense cedar; CECU, *Ceanothus cuneatus*; FECA, *Festuca californica*; FEID, *Festuca idahoensis*; FRCA, *Frangula californica* ssp. *occidentalis*; JUOC, western juniper; LIDE, tanoak; LIDEE, *Notholithocarpus densiflorus* var. *echinoides*; MEIM, *Melica imperfecta*; PICO, lodgepole pine; PIJE, Jeffrey pine; PILA, sugar pine; PIMO. western white pine; PSME, Douglas-fir; QUGAB, *Q. garryana* var. *fruticosa*; QUVA, *Q. vaccinifolia*; RHMA, *Rhododendron macrophyllum;* VAOV, *Vaccinium ovatum*; VAPA *Vaccinium parvifolium*; and XETE, *Xerophyllum tenax*.

9C3 Other cool landscapes

The vegetation distributions on cool soils with annual precipitation (MAP) >1500 mm differ greatly from those in the drier RCT. Large areas of more moist ultramafic soils have been mapped on the Josephine ophiolite (Fig. 9-5) and coast-ward from the Klamath Mountains across the Coast Ranges to Nesika Beach (Borine 1983, Fillmore 2005); and the distributions of plants have been recorded and evaluated on and near the Josephine ophiolite by Atzet et al. (!996) and by Jimerson et al. (1995).

The ultramafic soils on and around the Josephine ophiolite are mostly moderately deep and shallow Alfisols and Inceptisols, and deep Inceptisols in colluvium. Soil pH is generally slightly acid. The vegetation is mainly open conifer forest and shrubs. The sequence of trees and shrubs on cool ultramafic soils from the Josephine ophiolite toward the coast is represented by habitats E through I in Table 9.5. Besides the more common Douglas-fir, western white pine, tanoak, and bay trees, Port Orford cedar (*Chamaecyparis lawsoniana*) is commonly present in the wetter forests, and chinquapin (*Castanopsis chrysophylla*) is present in many areas. The main ferns are swordfern (*Polystichum munitum*) and bracken fern (*Pteridium aquilinum*).

Jimerson et al. (1995) reported the relative abundances of plant species on cool soils in ultramafic landscapes on the Six Rivers National Forest. The main shrubs, from more to less common, were huckleberry oak (*Q. vaccinifolia*), pinemat *manzanita* (*A. nevadensis*), red huckleberry (*V. parvifolium*), Pacific rhododendron (*R. macrophyllum*), greenleaf manzanita (*A. patula*), prostrate and dwarf ceanothus (*C. prostratus* and *C. pumilus*), oceanspray (*Holodiscus discolor*), California coffee-berry (*F. californica* ssp. *occidentalis*), Oregon-grape (*Berberis* spp.), serviceberry (*A. alnifolia*), serpentine silktassel (*G. buxifolia*), wood rose (*Rosa gymnocarpa*), and Sadler oak (*Q. sadleriana*). The more common, but mostly sparse forbs, along with beargrass (*Xerophyllum tenax*), were pipsissiwa (*Chimaphila umbellata*), iris (*Iris* spp.), modesty flower (*Whipplea modesta*), wintergreen (*Pyrola picta*), starflower (*Trientalis borealis* ssp *latifolia*), and rattlesnake plantain (*Goodyera oblongifolia*).

In order to get a perspective on the rate of ultramafic soil development, a sequence of soils on four recent to Pleistocene ultramafic outwash terraces were sampled where Rough and Ready Creek (RRC) flows from the Josephine ophiolite into the Illinois Valley (Table 9.6, Alexander et al. 2007b). The altitude is about 475 m and the current MAP is about 1500 mm. Pleistocene glaciation was confined to the upper part of the RRC watershed, but outwash was deposited where water from melting glaciers left the mountains (Hershey 1903). Gravels in the terraces were about 80 to 95% peridotite and serpentinite (Alexander et al. 2007b), and Si and Mg were dominant elements in the stream sediments (Miller et al. 1998). Spring and stream waters were characterized as Mg-bicarbonate water of pH 7.6 to 8.6, with a pH 8.2 mean (n=12) in September 1997. Concentrations of Ni from the 12 sample locations were 11 to 36 μg/L (Miller et al. 1998), which is very high compared to surface waters in nonultramafic terrain. The likely ages for the three outwash terraces were assumed to be about 15 ka, from the end of the last glaciation, 125 ka, from the end of the previous major glaciation, and much older for the highest terrace, probably on the order of 1000 ka (a million years). The soil on the recent alluvium was an extremely cobbly sandy loam, a slightly acid Mollisol (Pachic Haploxeroll), with a Jeffrey pine/Oregon white oak/birchleaf mountain mahogany plant community The soil on the 15 ka terrace was an extremely cobbly sandy loam (subsoil), an Inceptisol (Typic Haploxerept), with a Jeffrey pine/whiteleaf manzanita–dwarf ceanothus (*C. pumilus*) plant community. The soil on the 125 ka terrace was a very cobbly clay loam (subsoil), an Alfisol (Ultic Haploxeralf), with a Douglas-fir/whiteleaf manzanita–serpentine silktassel (*Garrya buxifolia*) plant community. And the soil on the 1000 ka terrace was a cobbly clay loam over cobbly clay, an Alfisol (Ultic Haploxeralf) with a kandic horizon, and a ponderosa pine–Douglas-fir–sugar pine plant community. The pH was strongly acid in the surface horizons of the oldest soil and moderately to slightly acid in the subsoil. Tree heights expected at 300 years based on growth curves (Dunning 1942) increase from 22 m on the lowest terrace to 55 m on the oldest terrace. Both the replacement of Jeffrey pine by ponderosa pine on the oldest terrace and good timber production indicate how ultramafic soils with small amounts of nonultramafic parent material can become highly productive when they are weathered and leached for millennia.

Table 9.6 Rough and Ready Creek ultramafic terraces. A comparison of B horizons, pedon available water capacity (AWC), and timber site index (TSI, tree height in meters at 300 years, Dunning 1942).

Terrace - Hor.	Depth (cm)	H:L:M proportions	free iron (Fe_d, %)	Munsell color (moist)	Exch. Ca/Mg	pH DW	AWC (cm/m)	TSI (m)
1-Bw	22-45	26:65:09	21	10YR 3/3	0.14	6.5	6	23
2-Bw	16-45	56:30:14	41	7.5YR 5/6	0.18	6.5	4	22
3-Bt	13-48	38:45:17	69	7.5YR 5/6	0.16	6.3	10	32
4-Bt1	24-42	56:22:22	119	5YR 5/6	0.67	5.6	12	55
4-2Bt2	42-60	-----	166	5YR 5/6	—	5.7	—	—

H:L:M proportions are the relative weights of heavy nonmagnetic (H), light (L), and heavy magnetic (M) grains in the fine sand (0.125-0.25 mm) separates by bromoform (SG=2.89) flotation and a hand magnet.

9C4 Cold and very cold landscapes, subalpine to alpine

Most of the cold and very cold soils of the Klamath Mountains are in, or near, glaciated terrains, and there are some ultramafic soils in many of the glaciated areas. The largest glaciated areas are centered around the highest peaks, which are Thompson Peak (2744 m asl) in the Trinity Alps and

Mt. Eddy (2754 m asl) on the Trinity ophiolite. Glaciated rocks around Mt Eddy are nearly all ultramafic rocks and gabbro, but those around Thompson Peak are mostly silicic granitic rocks, with some glaciated ultramafic rocks on the east side of the Trinity Alps. Inland, only mountains higher than about 2100 m asl had glaciers on them (Sharp 1960, Wahrhaftig and Birman 1965), but nearer the Pacific coast Ship Mountain (1556 m asl) had glaciers on it, particularly north from Brother No. 2. Terminal moraines from the larger glaciers in the Trinity Alps are as low as 750 m asl, much lower than the cirques. Inland the transition from cool to cold soils is on the order of 1500 to 1600 m asl– lower on N-facing slopes and higher on S-facing slopes. The cool to cold soil transition is much lower in coastal areas than inland.

Much of the area of very cold ultramafic soils (cryic STRs) is along mountain crests from China Peak (2604 m asl), which is about 12 km NW of Mount Eddy, southwestward tens of kilometers along the crest of the Scott Mountains and south from Mount Eddy a relatively short distance. The very cold soils, along with substantial rock outcrop and rubble, are lithic Entisols, deeper Entisols in colluvium, and moderately deep Inceptisols (Lanspa 1993). Above about 2400 m on Mt. Eddy, Whipple and Cole (1979) identified a *Penstemon procerus/Potentilla glandulosa* plant association with a higher elevation *Lesquerella occidentalis* (bladderpod) phase and a lower elevation *Eriogonum siskiyouense* (Siskiyou buckwheat) phase. Whitebark pine (*P. albicaulis*) grows at the highest elevations for trees, a short distance below the summit of Mount Eddy. Western white pine, red fir, and foxtail pine are trees that inland are exclusively in subalpine to alpine forests, although white pine also grows at lower elevations nearer the Pacific coast where the climate is wetter. Leatherleaf mountain mahogany (*Cercocarpus ledifolius*) is common on China Mountain and along many of the high ridges.

Figure 9-11. Cement Bluff. Gray till (G) below the yellow line is predominantly serpentinite, and brown till (B) above the yellow line is predominantly peridotite, partially serpentinized.

The cold ultramafic soils are mostly shallow to deep Inceptisols over bedrock and in glacial till and colluvium. Jeffrey pine, Douglas-fir, incense cedar, and white pine are common trees. Inland at the lower elevations of cold soils, sugar pine replaces western white pine. Some of the more common shrubs are huckleberry oak, greenleaf manzanita, and pinemat manzanita.

A large area of cold to very cold soils in ultramafic terrain, with minor gabbro, is shown in Figure 4-4A. The view is across High. Camp basin to Mount Eddy, with Cement Bluff (CB) visible on the far side of High Camp basin. Cement Bluff is a spectacular example of cemented till. Silica cementation is common in ultramafic glacial till and stream deposits of the Klamath Mountains (Alexander 1995). Cement Bluff is more than 100 m high, with brown peridotite till above and gray serpentinite till below (Fig. 9-11). A white deposit of nesquehonite and calcite persists on recessed surfaces where rain water does not reach them.

Mohr et al.(2000) have examined sediments in Bluff Lake (Fig. 4-4, 9-11), at the foot of Cement Bluff, that are as much as 13,500 years old. Two parallel ridges on the downslope side of Bluff Lake have serpentinite diamicton on the ends next to Cement Bluff and peridotite diamicton on the ends that are farther from Cement Bluff (Alexander 2004). Diamicton is a term used here, because it is uncertain whether the ridge deposits are glacial till or pronival rampart deposits, as described by Shakesby (1997). Minimal soil development on the ridges and their linear forms, rather than arcs, favor the pronival rampart hypothesis, which would allow ages <13,500 years for the ridge deposits.

Table 9-7. Bluff Lake ridges: properties of cambic (Bw) horizons in representative soils.

Soil Parent Material	Late (Younger) Bluff Lake Ridges				Early (Older) Bluff Lake Ridges			
	Bw (cm)	H:L:M portions	Color moist	pH DW	Bw (cm)	H:L:M proport.	Color moist	pH DW
peridotite	9-30	57:36:07	7.5YR 5/4	6.1	12-33	58:36:06	5YR 4/5	—
serpentinite	disc.*	20:68:12	10YR 5/3	6.5	12-36	28:59:13	10YR 4/3	6.4

* A discontinuous cambic horizon. Soil textures are all very cobbly loam. The heavy, light, and magnetic (H:L:M) proportions of fine sand grains are averages of four samples from ridge crest to lower slope positions. Minerals are mostly olivine and pyroxenes in the heavy nonmagnetic fractions and serpentine in the light fractions

The redder cambic horizons in the peridotite soils and the lack of continuous cambic horizons in serpentinite soils on the younger ridge (Table 9.7) indicate more advanced weathering and soil formation on the peridotite ends than on the serpentinite ends of the ridges. White fir trees are common on colluvium in the shadow of Cement Bluff, but not on the diamicton of the ridges. Lodgepole pine and azalea (*Rhododendron occidentale*) are common around Bluff Lake. Three ferns that are more common on ultramafics than on other materials are present in colluvium from Cement Bluff: *Adiantum aleuticum*, *Polystichum lemmonii*, and *Aspidota densa.* The most common forbs on the colluvium are *Anemone drummondii*, *Potentilla glandulosa*, *Sedum obtusatum*, *Galium glabrescens*, and *Pseudostellaria jamesiana*. Besides the huckleberry oak and manzanitas that are common on cold ultramafic soils in the High Camp basin, low juniper (*J.*

communis), Nootka rose (*Rosa nutkana*), and mountain spiraea (*S. splendens*), along with beargrass (*X. tenax*), are common in open conifer forests on the rocky ultramafic soils west of Bluff Lake.

Ship Mountain, only 30 km from the Pacific Coast, is one of the lowest Klamath mountains to have been glaciated during the last major stage of glaciation. Four Brothers, aligned from north to south, comprise Ship Mountain. All are near 1600 m high, with the highest at 1616 m asl (41.73°N, 123.80°W). There were several small glaciers on Ship Mountain, with the largest about four km long, from Brother No. 2 down toward the northeast to about 1000 m asl (Fig. 9-12). High precipitation as snow must have fed the Pleistocene glaciers to allow them to thrive at a low elevation. Currently the mean annual precipitation is >3,000 mm, but summers are dry, and cool. An extremely stony and very acid Inceptisol described on ground moraine at 1325 m asl near Jones Creek (site ShM2) had a depauperate forest of Douglas-fir, white pine (*P. monticola*), Brewer's spruce (*P. breweriana*), and sparse Lawson's cypress (*Chamaecyparis lawsoniana*). Shrubs in the forest were plentiful huckleberry oak (*Q. vaccinifolia*), scrub tanoak (*N. densiflorus*

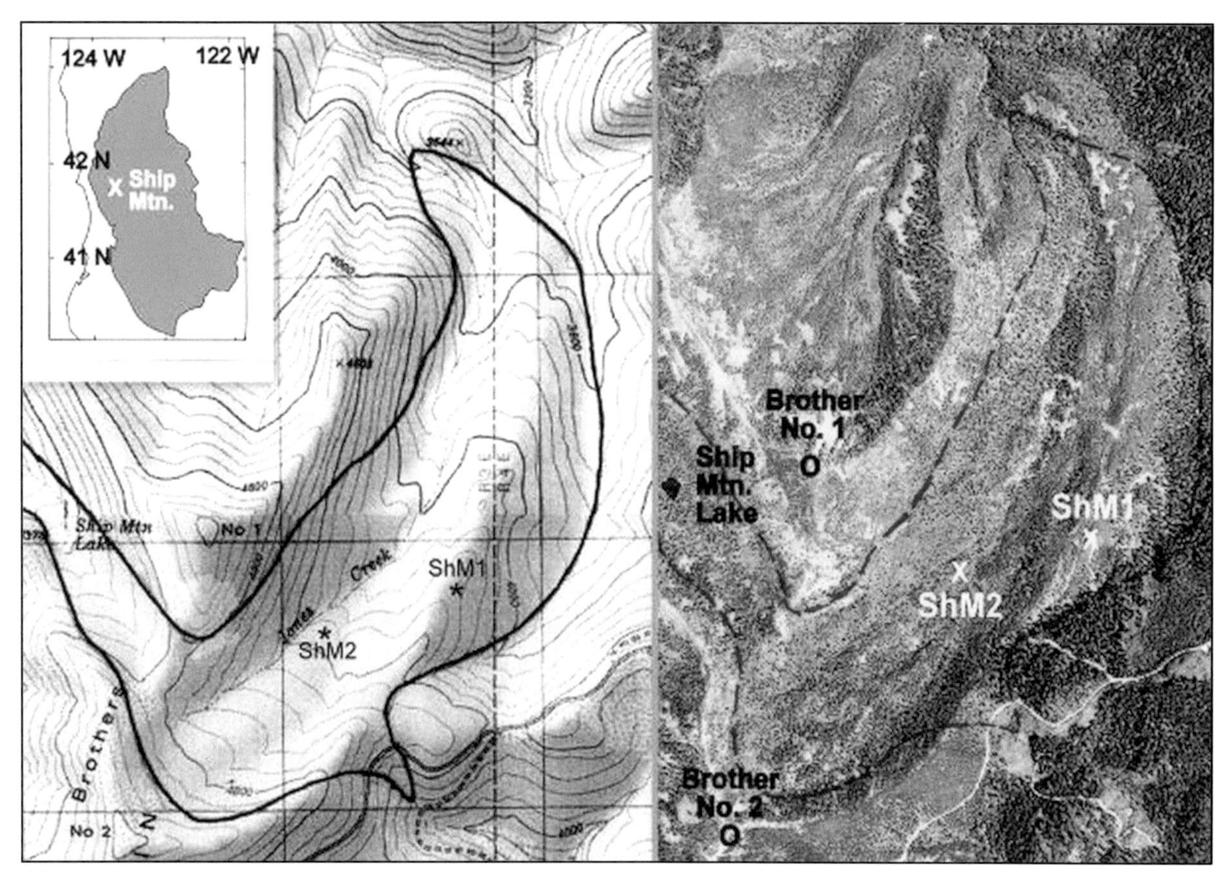

Figure 9-12. Ship Mountain glaciation: a topographic map and an aerial photograph. The delineation on the photograph (red dashes) shows the possibility that a branch from the large glacier down along Jones Creek might have flowed across the divide between Brothers 1 and 2 and beyond Ship Mountain Lake. Distances between the section lines on the topographic map are 1 mile (1.6 km).

var. *echinoides*), and red huckleberry (*V. parvifolium*), and sparse service berry (*A. alnifolia*) and greenleaf manzanita (*A. patula*), with common teaberry (*Gaultheria ovatifolia*), graminoids, and sparse beargrass (*Xerophyllum tenax*) at ground level, and with mosses on the forest floor. Another soil was described where the glacier passed over a ridge (41.747°N, 123.774°W) and deposited till of very cobbly loam, or very cobbly clay loam (site ShM1, Fig. 9-12). The soil there was a strongly acid over moderately acid Inceptisol with Oi/Oe/A/Bw/C horizons and open forest with lodgepole and white pine in the overstory, and with plentiful scrub tanoak and red huckleberry, common pinemat manzanita and huckleberry oak, and sparse serviceberry and greenleaf manzanita in the understory, and with common teaberry and sparse beargrass and mosses on the ground floor. All stones on the ground moraine were serpentinized peridotite, but the till on the ridge contained some pebbles and cobbles of gabbro and other nonultramafic rocks, but no more than 15% of them. Analyses of aqua regia digest from a Bw horizon (30-68 cm depth) sample showed 8.6% (86 g/kg) Fe, 0.16% (1.6 g/kg) Ni, and 11.5% (115 g/kg) Mg, with a Ca/Mg ratio of 0.02 g/g, and K was below the detection level of 0.01% (0.1 g/kg).

9C5 Serpentine fens

There are many different kinds of serpentine fens in the Klamath Mountains with sedges, rushes, bulrushes, grasses, and many different species of forbs. The fens are on cold to cool, wet Entisols,

Figure 9-13. Serpentine fens. A. A spring fed pitcher plant fen on the Trinity ophiolite, above Parks Creek. B. A stream side patch of pitcher plants in the High Camp basin; the hammer handle is 0.27 m (11 inches) long. C. The site where a Histosol was described in Wildmad fen. The shovel with its handle visible above the soil pit is 1.33 m (52 inches) long, and a Dutch auger was used to sample deeper.

Inceptisols, Mollisols, and Histosols. Some of the fens are limited to the margins of streams and ponds and some occupy small basins. Although the areas are small, the fens are important for their great habitat diversities. Most distinctive of the serpentine fens are those with pitcher plants (*Darlingtonia californica,* Fig. 9-13A, B). Pitcher plant fens, or bogs, are also found in wet depressions of nonultramafic sands behind beaches along the Pacific coast north of the Klamath Mountains.

The most well known serpentine fen is at the south foot (430 m asl) of Eight Dollar Mountain, on the Josephine ophiolite (Fig. 4-3). Botanists call it the TJ Howell Fen, after a late 19th century botanist (Ornduff 2008). The soil in the fen is a muck (a Haplosaprist), which means that it lacks the coarser fibers of peat. Although the soil in Eight Dollar fen (412 m asl) was not described in detail, a similar organic soil at 1190 m asl in the Wildmad (or Saddle Gulch) fen of the RCT was described in more detail (Fig. 9-13C). Eight Dollar fen occupies about 3 ha, and the wetter half of it is occupied by *D. californica*. Fen water on the upper end of Eight Dollar fen is neutral and it is slightly acid on the lower end of the fen.

Wildmad fen, where the soil was described, had neutral to slightly acid Oa horizons 9 and 27 cm thick and a 34 cm Oe horizon over neutral (pH 7) very dark gray silt loam at 70 to 80 cm and greenish gray silty clay loam layers of ultramafic sediment below 80 cm. In organic soils, Oa horizons consist of sapric material with low fiber contents, Oe horizons consist of hemic material with 17 to 40% fiber, and Oi horizons consist of fibric material with fiber >40%. The soil in the Wildmad fen is a Saprist, because the organic materials are predominantly sapric, and a Haplosaprist because there are no special features for this Saprist. Ultramafic sediment from a 70 cm depth to below 130 cm at the Wildmad site determines the subgroup classification to be a Terric Haplosaprist, and it is in a magnesic family. Plants on the fen are abundant graminoids (cottongrass, sedges (*Carex* spp.) and rushes (*Juncus* spp.), plentiful bistort (*Polygonum bistortoides*), and common *Parnassia californica*, *Symphyotrichum spathulatum*, *Hastingsia alba*, *Triantha occidentalis*, and tufted hairgrass (*Deschampsia cespitosa*). There is a fringe of azalea (*Rhododendron occidentale*) and sparse willowherb (*Epilobium oreganum*) around the margin of the fen. The absence of pitcher plants may be attributed to summer heat, although the Eight Dollar fen is at a lower elevation and might be warmer, or the absence of pitcher plants might be attributed to lack of plant dispersal. The Wildmad fen was referred to as the "Saddle Gulch Fen" by Sikes et al. (2013) They reported *Bryum pseudotriquetrum* as a common moss on the fen. A rare lady's slipper (*Cypripedium californicum*) was found in this fen and in other fens in the Klamath Mountains (Sikes et al. 2013).

9C6 Very old landscapes

During the Paleogene or early Neogene, Miocene terrains of the Klamath Mountains were eroded to a nearly level plain called the Klamath peneplain, and a practically contiguous Bell Springs peneplain crossed adjacent parts of the California Coast Ranges (Diller, 1902; Aalto, 2006). From the late Miocene, the Klamath Mountains and adjacent parts of the California Coast Ranges were uplifted episodically. Uplift was intermittent, allowing time for the development of broad valleys between episodes of uplift (Cater and Wells, 1953). Erosion has rounded most of the mountain summits and now the most strongly weathered soils are commonly on benches below the summits. The most strongly weathered soils are assumed to have developed during or following the Miocene. Strongly weathered ultramafic soils in the Klamath Mountains support noncommercial forests (Fig.3-5A), and they lack cultivation.

Early Cenozoic warm temperatures declined between the Eocene and Oligocene and then increased gradually though the Oligocene and early Miocene to means about 4° C greater than those of today during the Miocene. A middle Miocene decline reduced mean annual temperatures nearly to those of today. Then about 2.6 Ma ago, temperatures began to fluctuate widely through many cycles of glaciation. Thus, old Neogene soils on the Klamath and Bellsprings peneplains have not developed in tropical climates, yet some of these old Klamath Mountains ultramafic soils have characteristics that are generally unique to tropical soils (Hotz 1964, Alexander 2010, 2014a). In contrast to old tropical ultramafic soils, however, the very red (5YR to 10R) colors in most old ultramafic soils of the Klamath Mountains occur only in the upper meter, or half a meter, of the soils (Alexander 2010, Fig. 9-14). even in soils with Fe-oxide (FeOOH) contents >35 or 40% to depths greater than one meter.

Figure 9-14. The very old "Littlered" soil (a fine, ferritic, mesic Xeric Kandihumult) on Red Mtn., near Leggett CA.

As nontropical soils, the most unique features of old Klamath Mountains ultramafic soils are very high Fe concentrations and net positive charges that are related to high Fe-oxide contents. Other than these soils, the only North American nontropical soils with net positive charges at ambient pH are in volcanic materials where weathering produces amorphous aluminosilicates. Young ultramafic soils and those in dry climates have clay minerals with mostly negative surface charges. Weathering of ultramafic materials in humid climates of the Central Pacific Cordillera for thousands of years produces clays with Fe-oxides that have highly variable surface charges. Old, acid ultramafic soils with much Fe can have clays with more positive than negative surface charges. The anion-exchange capacities (AECs) in the old, weathered soils can be greater than the CECs.

The presence of net positive or negative charges in soils can be assessed qualitatively by pH differences in distilled water and in neutral salt solutions (Alexander 2014a). With negatively charged clays, cations in neutral salt solutions displace hydrogen ions and lower the pH in the solutions. In soils with positively charged clays, the anions in neutral salt solutions (for example Cl in KCl) displace hydroxyl ions and raise the pH in the solution. Chemists have more sophisticated explanations, but this less professional explanation is much easier for nonchemists to understand. The variable charges between pH 7.0 and pH 8.2 (pH 8.2 CEC minus pH 7.0 CEC) in old ultramafic soils were found to be 400 mmol+/kg (40 me/100 g), or greater; the CEC was about four times greater for these soils at pH 8.2 than for the soils at pH 7 (Alexander 2010). The large variable charge is attributable predominantly to "free" Fe-oxides, mainly goethite.

Besides the high Fe contents, all first transition elements from V to Ni and Cu and Zn are concentrated in old ultramafic soils as the Si and Mg are released from weathering ultramafic materials and removed by leaching (Hotz 1964, Foose 1992, Alexander 2014a). Weathering and leaching have concentrated nickel enough to support mining on Nickel Mountain (Pecora and Hobbs 1942, located at latitude 43°N in Fig. 9-5).

The depauperate forest on Gasquet Mountain (Fig. 3.5A), is representative of physiognomic aspects of vegetation on old ultramafic soils of the Klamath peneplain. Plants in the forest are very common western white pine and lodgepole pine, sparse Jeffrey pine and Douglas-fir, plentiful scrub tanoak (*N. densiflorus* var. *echinoides*) and red huckleberry (*V. parvifolium*), common coffeeberry (*F. californica* ssp. *occidentalis*), rhododendron *(R. macrophyllum*), low juniper (*J. communis*), and pinemat manzanita (*A. nevadensis*), sparse spiraea (*Spirea* sp.) and dwarf ceanothus (*C. pumilus*), common beargrass (*X. tenax*), and sparse forbs.

D Blue Mountains

There is a wide range of plant communities on ultramafic rock exposures in the Blue Mountains. They range from below 1000 m along the Snake River to about 2600 m asl (above-sea-level) in the mountains; they are mostly between 1050 and 2400 m asl. Currently, winters are cold and summers are dry (Fig. 9-2, Austin). Monthly air temperature means range from less than 0°C in January to more than 15°C in July. Mean annual precipitation ranges from about 400 mm at lower altitudes to 800 mm at higher altitudes.

Ultramafic soils in the Blue Mountains are similar to some of the cool to very cold ultramafic soils in the Klamath Mountains and the Sierra Nevada provinces. The ultramafic soils are commonly shallow to moderately deep, but deep soils are well represented, also. In the Blue Mountains, the ultramafic soils are Inceptisols, Alfisol, Mollisols, and Entisols.

Many of the ultramafic plants that are common in the Sierra Nevada and Klamath Mountains are common in the Blue Mountains, also. The vegetation ranges from sagebrush and grassland to conifer forests and alpine fell-fields. Among the more notable species differences are the absence of Jeffrey pine in the Blue Mountains, and grand fir (*A. grandis*) replaces white fir (*A. concolor*).

9D1 Aldrich Mountains and Strawberry Range

Six major serpentine vegetation types were recognized from about 1400 m in foothills adjacent to the valley of the John Day River up to the 2243 m summit of Baldy Mountain (Fig. 9-16) in the Strawberry Range (Table 9-8). They are represented by juniper woodland on Mollisols, conifer forests on Mollisols and Alfisols, and alpine fell-fields on very cold Mollisols (Tables 9-8 and 9-9). Soils were sampled in two of the vegetation types in the Aldrich Mountains and four were sampled from the foothills above the John Day Valley up to Baldy Mountain.

From low to high elevations, trees appear in the order western juniper, ponderosa pine, Douglas-fir, and grand fir. Lodgepole pine is present, or dominant, at the higher elevations in areas that have been disturbed or burned. Mountain mahogany (*Cercocarpus ledifolius*) is present in rocky areas at all elevations below the alpine fell-fields. Shrubs appear in the order sagebrush (*A. tridentata*), grouseberry (*Vaccinium scoparium*), pinemat manzanita (*A. nevadensis*), and, in the alpine fell-fields, buckwheat (*Eriogonum* spp.) subshrubs. Grasses are mainly bluegrass (*P. secunda*), Idaho fescue (*F. idahoensis*), and wheatgrass (*Pseudoroegneria spicata*), with wheatgrass dominant only at the lowest elevations. An interesting distribution at site A1 was the dominance of bluegrass under the south side of a juniper tree and the dominance of Idaho fescue under the north side. Elk sedge (*C. geyeri*) is present in forests at all elevations below the alpine fell-fields. No single forb species is dominant; yarrow (*Achillea millefolium*) and lanceleaf sedum (*S. lanceolatum*) are ubiquitous forbs and lupines (*Lupinus* spp.) are widespread. Some of the more common forbs in the alpine fell-fields are wavewing (*Pteryxia terebinthina*), chickweed (*Cerastium arvense*),

Figure 9-15. The summit of Baldy Mountain in the Strawberry Range of the Blue Mountains.

cushion draba (*D. densiflora*), frasera (*F. albicaulis*), fringed-onion (*Allium fibrillum*), balsam root (*Balsamorhiza serrata*), buttercup (*Ranunculus* spp.), scabland fleabane *(Erigeron bloomberg*), paintbrush (*Castilleja* spp.), and alpine pennycress (*Noccaea fendleri*). Some varieties of alpine pennycress are nickel hyperaccumulators. Boulders on Baldy Mountain are adorned with yellow to orange (*Caloplaca* sp.) and light green (*Lecanora* sp.) lichens.

Table 9.8 Ultramafic soils and dominant plants in the overstory/shrub/herb layers of plant communities (PCs) in an elevation sequence on the north sides of the Aldrich Mountains and Strawberry Range. The sequence is presented in the order of increasing precipitation.

PC	Alt. (m)	MAP (mm)	Soils and the More Common Plants
A1	1420	400	lithic, argillic Mollisol, Argixeroll: JUOC/FEID–AGSP–POSE
A2	1370	450	lithic, haplic Mollisol, Haploxeroll: JUOC/CELE/POSE–FEID
B	not sampled		ponderosa pine (PIPO) forest with cool (mesic STR) Mollisols
C	1480	500	argillic Mollisol, Argixeroll: PIPO-PSME/AMAL/FEID-CAGY
D	1640	600	Alfisol, Haploxeralf: PSME–PIPO–ABGR/VASC–ARNE/CAGY
E	1875	700	Alfisol, Haploxeralf: PSME–PIPO–ABGR–PICO/ARNE–VASC/CAGY
F	2200	800	lithic, haplic Mollisol, Haplocryoll: POSE–FEID and many forb speciscoulorumes

The plant species symbols are ABGR, *Abies grandis*; AGSP, *Pseudoregnaria spicata*; AMAL, *Amelanchier alnifolia*; ARNE, *Arctostaphylos nevadensis*; CAGY, *Carex geyeri*; CELE, *Cercocarpus ledifolius*; FEID, *Festuca idahoensis*; JUOC, *Juniperus occidentalis*; PICO, *Pinus contorta*; PIPO, *Pinus ponderosa*; POSE, *Poa secunda*; PSME, *Pseudotsuga douglasii*; VASC, *Vaccinium scoparium*.

Kruckeberg (1969) visited several locations in the Aldrich Mountains and Strawberry Range. Recently, plant lists (unpublished) have been compiled by US Forest Service personnel.

Peridotite lacking serpentinization was found on Baldy Mountain, and elsewhere in the Strawberry Range, but the dominance of light minerals (L in column 5 of Table 9.9) in the fine sand fractions of the soils in Tables 9.8 and 9.9, indicate that the soil parent materials are highly serpentinzed peridotite. The light minerals are mostly serpentine and some chlorite. The soils have very little total Ca, as indicated by the mass Ca/Mg (g/g) ratios in Table 9.9. Much of the Ca is exchangeable and available to plants, however, as indicated by exchangeable Ca/Mg ratios. Nevertheless, the exchangeable Ca/Mg ratios are still quite low compared to nonultramafic soils, and low Ca availability is likely a major reason that the plant distributions are unique on the ultramafic soils..

Table 9.9 Subsoil horizons in soils of the plant communities in Table 9.8. Iron and nickel are from aqua regia digestion (percent is g/kg divided by ten).

PC	Soil Hor.	Depth (cm)	Texture (feel)	H:L:M[a] fine sand	Munsell color (m)	Fe	Ni	Ca/Mg ratio[b]		pH DW
						g/kg		g/g	molar	
A1	Bt1	12-18	vGrCL	17:81:02	7,5YR 3/3	64.4	1.61	0.02	0.18	6.1
A2	A3	7-17	GrSL	02:96:02	10YR 3/2	53.2	1.51	0.02	0.24	6.2
C	Bt	28-42	vGrCL	26:72:02	10YR 4/4	–	–	–	0.32	5.8
D	Bt	12-45	vGrCL	16:81:03	5YR 4/4	63.5	1,28	0.06	0.24	5.6
E	Bt	38-68	vGrCL	04:95:01	7.5YR 4/4	94.0	2.99	0.02	0.13	6.0
F	A2	5-26	GrL	02:94:04	10YR 3/2	56.7	1.32	0.03	0.40	5.9
F’	Bw	28-48	GrL	09:90:01	10YR 3/4	48.3	1.30	0.03	0.11	6.4

[a] H:L:M proportions are the relative weights of heavy nonmagnetic (H), light (L), and heavy magnetic (M) grains in the fine sand (0.125-0.25 mm) separates by bromoform (SG=2.89) flotation and a hand magnet.
[b] Ca/Mg ratios based on aqua regia digestion (g/g) and on KCl or NH_4Cl extracted cations (CEC, molar).
The soil of plant community F that is labeled F’ is a haplic Mollisol in colluvium on Baldy Mountain.

9D2 Vinegar Hill

There are some moderately large areas of ultramafic rocks around Vinegar Hill. A few very small ultramafic areas with alpine vegetation are present up to about 2320 m on the south side of a rocky ridge that trends northwest from Vinegar Hill. The north slope from the ridge has been glaciated, but not the south slope. Upper convex slopes of the ridge are rocky and have very shallow to shallow Mollisols (L. Haplocryolls) and Inceptisols (Eutrocryepts). Southward from the ridge the characteristic soils are moderately deep Alfisols (Mollic Haplocryalfs) on nearly linear “backslopes” and deep Mollisols on lower concave slopes. Fine sand from a surface soil sample was found to contain appreciable glass and nonultramafic minerals, in addition to ultramafic minerals, indicating the influence of volcanic ash. Sparse whitebark pine (*P. albicalis*) and low juniper (*J. communis*) are present along the rocky ridge, and patches of stunted subalpine fir (*A. lasiocarpa*) forest occur on nonultramafic soils of the upper N-facing slope. On the south, a shallow ultramafic Mollisol (Haplocryoll) at about 2310 m asl on an upper convex slope had sparse whitebark pine trees, common sagebrush (*A. tridentata*) and subshrubs, common sedges, and many species of forbs. A moderately deep Alfisol (Haplocryalf) at about 2290 m asl near the inflection point from the upper convex to a lower concave slope had sparse ponderosa pine trees, very common sagebrush and subshrubs (cushion phlox (*P. pulvinata*) and buckwheat), plentiful sedges (*Carex rossii* and *C. geyeri*), sparse blue wild rye (*Elymus glaucus*), and common forbs. The lower, concave slopes were dominated by sagebrush (*A. tridentata*) and mountain sorrel

(*Rumex* sp.), with patches of buckwheat (*Eriogonum flavum* and *E. heracleoides*).

9D3 Glasgow Butte

The ultramafic rocks in the Glasgow Butte area are about 12 km south-southwest of Sparta. They are present in a small area of a few hundred hectares on the south side of Glasgow Butte, which is south of the Wallowa Mountains (Prostka 1962). Exposures of peridotite and serpentinite range from about 1050 m asl in hilly terrain on the south side of the butte up nearly to the top of Glasgow Butte. The summit of Glasgow Butte at about 1500 m is basalt. Mean annual precipitation is about 200 to 300 mm.

Mollisols (Argixerolls) prevail on both peridotite and serpentinite in the Glasgow Butte area. The dominant vegetation is sagebrush (*Artemisia tridentata*) with wheatgrass (*Pseudoroegneria spicata*) and bluegrass (*Poa secunda*). Sparse bitterbrush (*Purshia tridentata*) occurs on some slopes. Grazing by livestock is intensive and *Poa bulbosa* is common. Some juniper (*Juniperus occidentalis*) trees have been present, but practically all of them have been cut to deny raptors perches for spotting sage hens.

Only one ultramafic soil was described in this area. Although the bedrock beneath the described soil is serpentinite, pebbles and sand grains in the soil indicate a substantial contribution from basalt that once covered the serpentinite and has now been removed by geological erosion. The soil is a cool Mollisol (a Lithic Argixeroll).

9D4 Blue Ridge

Blue Ridge is on a highly dissected plateau south of the Aldrich Mountains. Serpentinized peridotite and shallow ultramafic soils are exposed along the summit of the ridge. A very shallow ultramafic Mollisol was described below the summit, at about 1540 asl. The vegetation is scanty, with juniper trees in patches of deeper (yet shallow) soils. Bluegrass is common, and under trees it is plentiful, with some Idaho fescue near the trunks of trees; wheatgrass is sparse. The most abundant forb when the soil was described in June, 2013, was Tolmei's onion (*A. tolmiei*), followed in abundance by much less common rough eyelash (*Blepharipappus scaber*).

10 Coastal Southwest

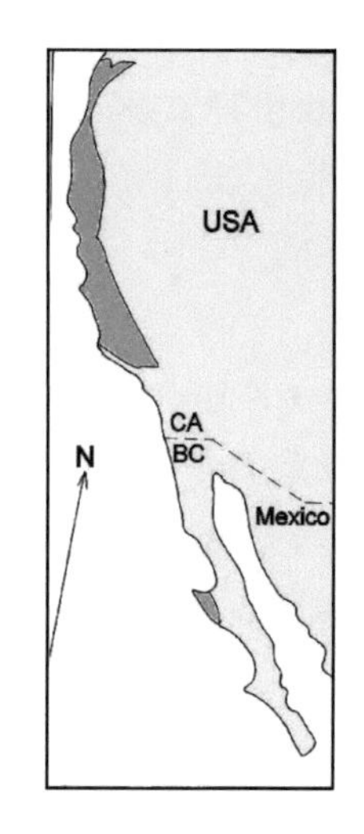

The California Coast Ranges, which extend up to about 43°N latitude in Oregon, contain oceanic terranes with ultramafic rocks, and there are similar terranes in Baja California. These terranes were added to North America from the Jurassic through much of the Cenozoic, until the Pacific plate ceased to be subducted beneath the continent. Currently the Pacific plate is moving north-northwestward west of the San Andreas fault. During the Neogene a block of north-northwest trending terranes lacking ultramafic rocks were rotated across the north-northwest trend of the oceanic coastal terranes (Nicholson et al.1994). This rotated block now comprises the core of western part of the east-west trending Transverse Ranges (Fig. 10-1A). North-northwest trending terranes south of the Transverse Ranges similar to those of the California Coast Ranges occur on the Vizcaíno Peninsula and on islands along the coast south from the Santa Cruz Island and Santa Rosa Island faults to Magdalena and Margarita Islands (Fig. 10), including Santa Catalina Island off the coast of southern California and Cedros Island north of the Vizcaíno Peninsula.

Most of the ultramafic rocks in the California Coast Ranges and Baja California are in moderately steep to very steep mountainous terrain. Climates range from cool and wet, with up to 3000 mm of precipitation annually

Figure 10-1.
A. California Coast Ranges. BB, Big Blue Hills; CC, Curry County; Ce, Cedars; CR, Coyote Ridge; CRO, Coast Range ophiolite; CuR, Cuesta Ridge; FM, Figueroa Mtn.; JR, Jasper Ridge; La, Lassics; MD, Mount Diablo; MT, Mt. Tamalpais; NI, New Idria; NM, Nickel Mountain; SAF, San Andreas fault; SCI, Santa Catalina Island; SCIf, Santa Cruz Is. fault; SF, San Francisco; TM, Table Mountain.
B. Baja California. Cam, Camalli; CI, Cedros Island; MdI, Magdalena Island; MgI, Margarita Island; mm, magnesite mine; VT, Vizcaino terrane.

west of the Klamath Mountains, to warm and summer-dry south of the Klamath Mountains (Napa, Fig. 10-2), and to hot and arid in Baja California (Santa Monica, Fig. 10-2). Soil temperature regimes range from cold (frigid) to hot (hyperthermic STR), with mostly cold and cool soils (frigid and mesic STRs) on the north and cool to warm soils (mesic and thermic STRs) farther south in the California Coast Ranges. All ultramafic soils in Baja California are hot and arid. The highest summit areas from the Yolla Bolly Mountains south to Snow Mountain have been glaciated, but no ultramafic rocks in the California Coast Ranges are known to have been glaciated during the Pleistocene. Soil moisture regimes of the ultramafic soils are mostly summer-dry (xeric SMR), moist (udic SMR) near the coast in the far north of the California Coast Ranges, and arid (aridic SMR) in Baja California.

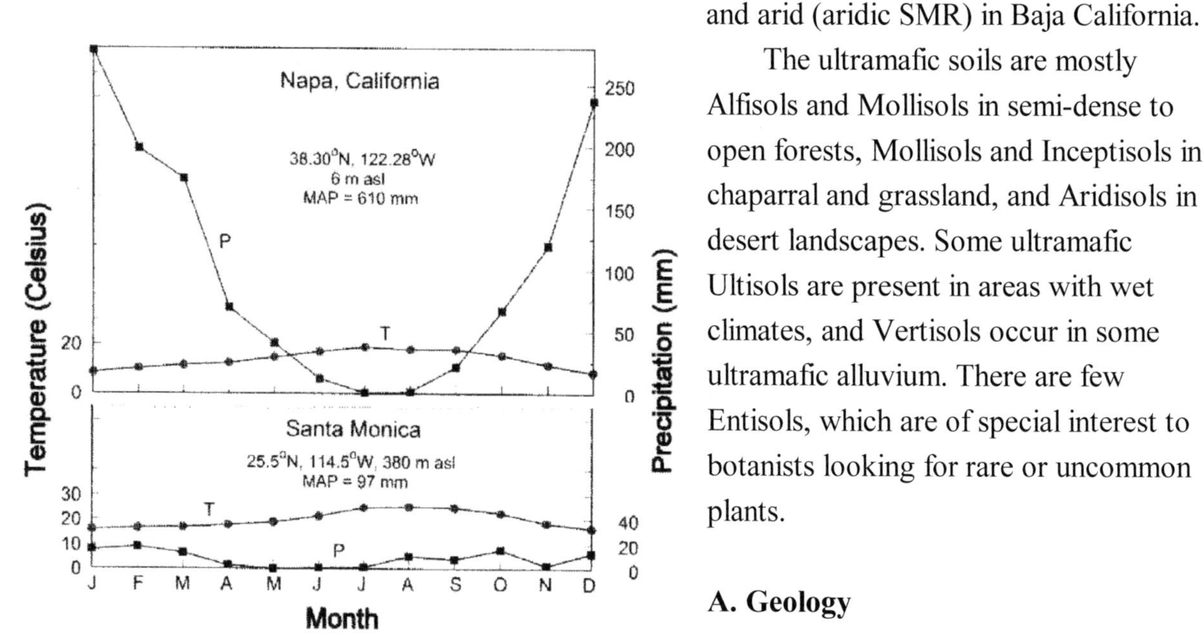

Figure 10-2. Mean monthly temperatures and precipitation at Napa CA and Santa Monica BC.

The ultramafic soils are mostly Alfisols and Mollisols in semi-dense to open forests, Mollisols and Inceptisols in chaparral and grassland, and Aridisols in desert landscapes. Some ultramafic Ultisols are present in areas with wet climates, and Vertisols occur in some ultramafic alluvium. There are few Entisols, which are of special interest to botanists looking for rare or uncommon plants.

A. Geology

As oceanic terranes drifted to North America from the Jurassic to the Paleogene, ocean sediments accumulated on them. Some ocean sediments were scraped off and deposited in trenches where oceanic crust was subducted as it approached the continent. Most of the subducted oceanic crust was returned to the mantle, but some of it was detached and eventually added to continental crust. Oceanic crust that was subducted to depths of about 15 to 30 km and returned to more shallow depths before it could be heated to the high temperatures at those 15-30 km depths was metamorphosed to the blueschist facies. Blocks of blueschist that are very hard and resistant to weathering (for example, Turtle Rock, Fig. 10) are common in mélange from Oregon to California and in Baja California.

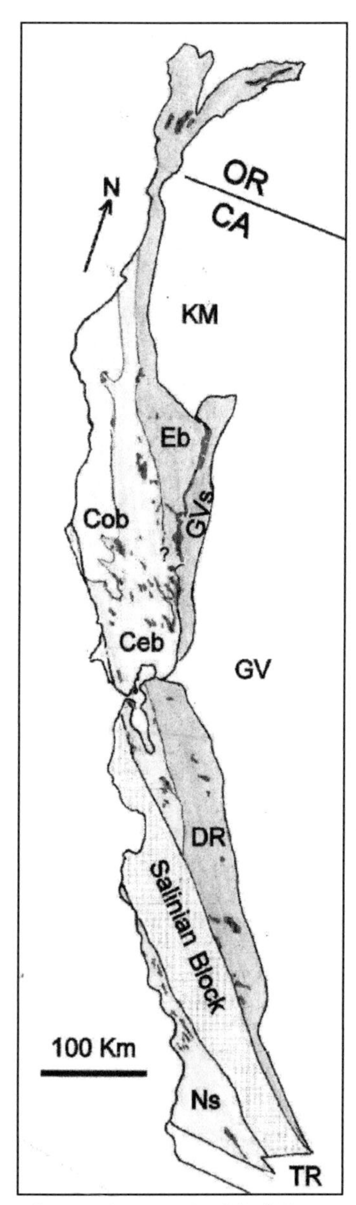

Figure 3-10. Ophiolitic terranes in the California Coast Ranges; Ceb, Central belt; Cob, Coastal belt; DR, Diablo Range; Eb, Eastern belt; GV, Great Valley; GVs, Great Valley sequence; KM, Klamath Mountains; Ns Nacmiento sector; TR, Transverse Ranges.

Rocks of oceanic origin in the California Coast Ranges that were once called the Franciscan formation in California and the Dothan formation in Oregon are now referred to as the Franciscan complex (Wakabayashi 2015). Mostly, they are dirty sandstone with lesser amounts of mudstone, or shale, and chert; limestone is much less common. The dirty, or muddy, sandstones, which are called *graywacke*, contain considerable silt and clay, along with the sand. Sands in the graywacke are feldspars, quartz, and fragments of basalt and chert. In contrast to clean sandstones that are not muddy and commonly weather to yield sandy soils, graywacke soils are generally not sandy. The abundance of chert and paucity of limestone implies that sedimentation was at depths >4000 m in the ocean, where the high pressure and low temperature inhibit the formation of lime ($CaCO_3$), which would crystalize to form calcite or aragonite. Peridotite in Baja California and the California Coast Ranges is generally serpentinized extensively and the basalt is commonly altered to greenstone.

B. California Coast Ranges

Robert Coleman has good descriptions of the geology, and maps with the locations of some ultramafic rocks in the Coast Ranges (Alexander et al. 2007a). Ultramafic rocks of the Coast Ranges are in the ophiolitic bodies shown in Figure 10-3 (Irwin 1977). The Ophiolitic bodies in California are shown in a map compiled by Churchill and Hill (2000). In northern California, Bailey et al. (1964) recognized three areas with different stages of alteration, or metamorphism, in the Franciscan complex, which have been called the Eastern, Central, and Coastal belts (Fig. 10-3). Tectonic and metamorphic alteration increase from the Coastal to the Central and Eastern belts. The Coastal belt lacks ophiolitic terranes. Farther south, the main belts of the California Coast Ranges are the Diablo Range and Nacimiento sector, separate\ed by the Salinian block of non-oceanic terranes that have been transported hundreds of kilometers from the south along the west side of the San Andreas fault (Fig. 10-3).

Blueschist metamorphism is common in mélanges of the Eastern and Central belts and in the Diablo Range and the Nacimiento sector, indicating that subduction of rocks to depths between 15 and 30 km has been common. There is even some eclogite, which implies an even greater depth of subduction.

Ultramafic rocks are most common in the Eastern belt and the Diablo Range (Fig. 10-3). Much of the ultramafic rock in the California Coast Ranges may have been emplaced by overthrusting from the Klamath Mountains and from the Coast Range ophiolite. The Coast Range ophiolite (COR) is exposed beneath the eastern margin of the Great Valley sequence of sediments (Fig. 10-3) and remnants of it are common as far coastward as Cuesta Ridge and Point Sal. The CRO has an obscure history (Wakabayashi 2015). It is a controversial subject. Both subducton and obduction could be involved in its origins (Orme and Surpless 2019).

10B1 Eastern belt

The Eastern belt (Fig. 10-3) consists largely of a stack of coherent thrust sheets of mainly metaclastic rocks (Wakabayashi 2015) in the Yolla Bolly, Pickett Peak, and other terranes. Graywacke and other marine sedimentary rocks prevail in the Yolla Bolly terrane, with some metavolcanic rocks and sparse blocks of amphibolite and blueschist (Blake and Jones 1981). The Pickett Peak terrane includes the South Fork Mountain schist in California and the Colbrooke schist in Oregon. Ultramafic rocks occur along the coast at Gold Beach and at Nesika Beach in Curry County, Oregon (CC, Fig. 10-1A).

Nickel Mountain (NM in Figure 10-1A). A Ni-laterite has developed in a no more than slightly serpentinized peridotite that has been thrust across the Canyonville fault from the Klamath Mountains onto the Dothan formation (Ramp 1978). The Ni-laterite is at about 800 to 1000 m asl on a mountain summit and on benches below the summit. The Ni- laterite has been described by Pecora and Hobbs (1942) and subsequently by Ramp (1978). Four zones of soil, or regolith, were described above parent rock: (1) dark reddish brown soil with 1-8 g Ni/kg; (2) yellowish brown soil with 6-15 g Ni/kg; (3) yellowish brown soil with some quartz-garnierite boxwork and 10-25 g Ni/kg; (4) weathered peridotite with quartz-garnierite boxwork in greenish brown saprolite with 5-15 g Ni/kg; and (5) parent rock with 2-3 g Ni/kg at depths >12 m.

The mean annual temperature at 200 m asl in Riddle, 7 km east-southeast of Nickel Mountain is 19°C and the precipitation is 775 mm, with only 8 mm in July. No natural soils were mapped on the drastically disturbed area of Ni-laterite, but moderately deep Alfisols (Dubakella series) and shallow Inceptisols (Pearsoll series) were mapped in adjacent areas of ultramafic rocks (Johnson et al. 2003). These are cool, summer-dry soils with semidense to open conifer forests and understories of shrubs, commonly buckbrush (*Ceanothus cuneatus*), and grasses.

Curry County. Curry County, Oregon, has many ultramafic soils in a wide range of climates, and it has a recently published detail (order 2) soil survey (Fillmore 2005). About 422,000 ha were mapped in the soil survey and 9% of the area was considered to have ultramafic (magnesic family) soils. Approximately 35% of the ultramafic soils are cold and summer-dry, 60% are cool and summer-dry, and 5% are cool and moist through summers (isomesic STR and udic SMR). A majority of the cold soils are in the Klamath Mountains and a majority of the cool soils are in the California Coast Ranges. Slope gradients are mostly steep, both in the mountains and on hills near the coast. Mean annual precipitation ranges from 1500 to 2000 mm along the coast to >3000 mm at higher elevations inland. Most of the precipitation is rainfall near the coast, and also snow in the higher mountains.

The ultramafic soils are mostly Alfisols and Inceptisols. Many of the soils are shallow, but more of them are moderately deep to deep, and many soils are very deep in colluvium. The soils with isomesic soil temperature regimes are mostly Mollisols. The soils are generally slightly acid.

The cool and cold ultramafic soils have semidense to open forests with shrubs in the understories. Douglas-fir, incense cedar, Jeffrey pine, western white pine, Port Orford cedar, and tanoak are the common trees. Knobcone pine (*P. attenuata*) is common in areas that have been burned severely. Huckleberry oak (*Q. vaccinifolia*), red huckleberry (*V. parvifolium*), Pacific rhododendron *(R. macrophyllum*), azalea (*Rhododendron occidentale*), serpentine silktassel (*Garrya buxifolia*), California coffeeberry (*Frangula californica* ssp. *occidentalis*), pinemat manzanita (*A. nevadensis*), prostrate ceanothus (C. *prostratus*), and beargrass (*Xerophyllum tenax*) are common in the understories. California bay (*Umbellularia californica*) is common as a shrub. Pipsissawa (*Chimaphila umbellata*) is the most ubiquitous and abundant forb in the forest understories. Greenleaf manzanita (*A. patula*) is common on the cold soils and Sadler oak (*Q. sadleriana*) is common locally. Also, evergreen huckleberry (*V. ovatum*) and oceanspray (*Holodiscus discolor*) are common on some cool soils. Savannas prevail on the cool moist soils, with sparse Douglas-fir and Port Orford cedar (*Chamaecyparis lawsoniana*), and with bentgrass (*Agrostis* sp.), velvet grass (*Holcus lanatus*), and other grasses. Swordfern (*Polystichum munitum*) and bracken (*Pteridium aquilinum* var. *lanuginosum*) are common ferns. Western red-cedar (*Thuja plicata*) and Sitka spruce (*Picea sitchensis*) are common near the coast. Sitka spruce is most common where ocean spray is blown inland.

10B2 Coast Range ophiolite

The Coast Range ophiolite (CRO, Fig. 10-1A) is exposed along the edge of the Great Valley, The ophiolite is stratigraphically below the Great Valley sequence of sedimentary strata (GVs, Fig. 10-3) that are exposed in hills east of the Coast Ranges. The COR contains gabbro, basalt, ultramafic rocks, and sedimentary rocks. It dates from a relatively short period, 172 to 161 Ma, during the Jurassic (Hopson et al. 2008). Its origin has been controversial (Dickinson et al. 1996). East of the Coast Ranges, the ophiolite is overlain by about 12 km of late Jurassic and Cretaceous sedimentary rocks of the Great Valley group. Relatively large areas of the ophiolite are exposed largely along the eastern margin of the California Coast Ranges south of the Klamath Mountains to Lake Berryessa, and along oblique trends from near Lake Berryessa northwestward into the Coast Ranges south of Clear Lake (Fig. 10-3). Some rocks of the Coast Range ophiolite have been thrust over the Franciscan complex as far as the Pacific coast. Ultramafic rocks in the ophiolite range from pristine peridotite and dunite to completely serpentinized peridotite and to dismembered and scrambled ophiolite and serpentine mélange (McLaughlin et al. 1988). Old deposits from earth flows or submarine slides of weak, or plastic serpentine mélange are common; they are called *olistostromes* (Phipps 1984).

Climates along the eastern margin of the Coast Ranges, south of the Klamath Mountains, are cool to mostly warm, and summer-dry. Dominant warm ultramafic soils on the steep to very steep mountain slopes are shallow Mollisols and Alfisols (Lithic Haploxerolls and Argixerolls (Fig. 1-4C and 10-4), as mapped by Gowans (1967), Begg (1968), and Reed (2006). Mostly, the vegetation is chamise and mixed chaparral with sparse gray pine trees. The main cool soils are moderately deep Alfisols (Haploxeralfs) with open forests that commonly have Jeffrey pine–incense cedar–Douglas-fir/buckbrush/Idaho fescue plant communities, and lithic Mollisols (Lithic Argixerolls). Todd Keeler-Wolf conducted a comprehensive investigation of the serpentine vegetation at Frenzel Creek (Keeler-Wolf 1983, 1990). The serpentine chaparral there is dominated by whiteleaf manzanita (*A. viscida*) and leather oak (*Q. durata*). Serpentine endemic, or practically endemic, McNab cypress (*Hesperocyparis macnabiana*) occurs in chamise chaparral, and Sargent cypress (*Hesperocyparis sargentii*) occurs in riparian areas along Frenzel Creek (Chang 2004).

Figure 10-4. Whiteleaf manzanita chaparral on Mollisols and Alfisols of the Coast Ranges ophiolite (serpentinized peridotite, right, and metagraywacke, left), with sparse gray pine on the metagraywacke..

Ultramafic sediments eroded from Walker Ridge have accumulated on alluvial fans on the west side of Bear Valley and as basin-fill in the valley. Soils on the ultramafic fans are predominantly Mollisols (Pachic Haploxerolls) and those on the basin-fill are Vertisols (Endoaquerts, Reed 2006). Bear Valley is well known by botanists for spectacular spring wildflower displays (Edwards 1994). Botanists favor grazing on the grassland of Bear Valley, because it can enhance wildflower displays by reducing competition from exotic grasses.

On exposures of the ophiolites northwest of Lake Berryessa the ultramafic soils and vegetation are similar to those along the eastern margin the Coast Ranges, but there are less of the very steep slopes and more moderate slopes and more serpentine grassland with lithic Mollisols (L. Haploxerolls and L. Argixerolls, Lambert et al. 1978, Smith and Broderson 1989). Leather oak and whiteleaf manzanita are common shrubs (Fig. 1-4C), with some chamise chaparral on warm south-facing slopes with very hot summers, and with some McNab cypress on cooler slopes. Other relatively common chaparral shrubs are Jepson ceanothus (*C. jepsonii*), silktassel (*Garrya* sp.), and toyon (*Heteromeles arbutifolia*) (Colwell et al. 1955). Gray pine (*P. sabiniana*) trees are sparse, and there are no blue oak (*Q. douglasii*) trees on the ultramafic soils.

Much of the McLaughlin Reserve managed by the University of California, Davis, has ultramafic rocks of the CRO. A substantial amount of ultramafic related ecological research has been conducted in and around the Reserve. Some of the publications from that research are about the population biology of rare plants (Jurjavcic et al. 2002, Wolf et al. 2000, Harrison et al. 2000), local and regional species diversity (Harrison 1997, 1999), fire ecology of serpentine chaparral (Safford and Harrison 2004), weeds and revegetation on ultramafics (Williamson and Harrison 2002), and grazing effects on ultramafics (Safford and Harrison 2001).

10B3 Central Belt

The common rocks in the Central Belt are graywacke and other marine sedimentary rocks, with lesser amounts of metavolcanic rocks and sparse blocks of amphibolite and blueschist. Ultramafic habitats are described for ultramafic rock exposures at the Lassics, the Cedars, Ring Mountain, Jasper Ridge, and Coyote Ridge.

Figure 10-5. Contrasts in cool to cold, summer-dry, ultramafic soils of the Lassics area, California. A. An Entisol with no plants other than sparse bedstraw (*Galium* sp.). B. A shallow, haplic Mollisol with few Jeffrey pine and incense cedar trees, huckleberry oak (*Q. vaccinifolia*), buckbrush *(C.cuneatus*), and sparse bedstraw (*G. ambiguum*). C. A moderately deep Alfisol with an open Jeffrey pine, incense cedar forest containing some white fir and sugar pine trees, pinemat manzanita (*A. nevadensis*) and serviceberry (*A. alnifolia*) and sparse *Phlox diffusa* and *Galium ambiguum*.

Lassics. Detail soil surveys (orders 1 and 2) were done on a small area to characterize the habitat of the Lassics lupine (*L. constancei*), (Alexander, 2008, *A Soil Survey of Serpentine Landscapes in the*

Lassics Area, unpublished manuscript, Six Rivers National Forest, Eureka, CA). Ultramafic rocks of the Lassics area were thrust from the Coast Range ophiolite, or the Pickett Peak terrane, to the Lassics in the Central Belt of the California Coast Ranges, where they now reside (Kaplan 1984).

Most of the ultramafic soils in the Lassics aea are Inceptisols and moderately deep Alfisols on moderately sloping to moderately steep terrain, with less soils on steep and very steep slopes. The landscapes range from open Jeffrey pine–incense cedar forest with buckbrush or pinemat manzanita on moderately deep to very deep, cool to cold Alfisols and Inceptisols to shallow Inceptisols, and practically barren, very shallow Entisols (Fig. 10-5). There is an entirely barren area (Fig. 4-2B) on a completely serpentinized harzburgite. The Forest Service of the USDA manages the Lassics area to protect a unique combination or rare plant species.

The Cedars. The Cedars area gets its name from groves of Sargent cypress on serpentinized peridotite in the Austin Creek watershed, which is tributary to the Russian River near where it empties into the Pacific Ocean. It is one of the first places where the occurrence of low temperature serpentinization was recognized by Ivan Barnes (Barnes et al. 1967). While Mg and Si from the hydrolytic alteration of olivine and pyroxenes are retained in the production of serpentine and other Mg minerals, Ca is incompatible (defined in the glossary) in those minerals and leaves the peridotite as a hydroxide (Chapter 2, 2F and 2G). Water containing the Ca-hydroxide is extremely alkaline (pH >11). Some of the Ca-hydroxide water surfaces in springs where the Ca reacts with carbon dioxide (CO_2) in the atmosphere or bicarbonate in surface water to produce Ca-carbonate, which is deposited as travertine (Fig. 2-8).

Figure 10-6. Roger Raiche showing Robert Coleman a very steep slope across Austin Creek.

The Cedars is a steep mountainous area with some very steep canyon sides with sparse soil and practically devoid of plants (Fig. 10-6). Elevations range from about 100 m along Big Austin Creek to about 600 m on the higher ridges. Mean annual precipitation is in the 1500 to 1600 mm range, with dry summers.

Ultramafic soils, other than very shallow Entisols, were mapped mostly as shallow Inceptisols and Mollisols, and some moderately deep to deep Alfisols (DeLapp and Smith 1978, quadrangles 61D-1, 2, 3, 4). The mapped soils are reddish, with 5YR to 2.5YR hues. The more barren areas have been mapped as rock outcrop, ignoring very shallow Entisols (Miller 1972).

The vegetative cover is mostly mixed chaparral and groves of Sargent cypress (*Hesperocyparis sargentii*) trees

(Raiche 2009). The main chaparral species are leather oak (*Q. durata*) and sticky and Cedars manzanitas *(A. viscida* ssp. *pulchella* and *A. bakeri* ssp. *sublaevis*); and chamise, Jepson ceanothus, buckbrush, toyon, and California bay are common shrubs. On north-facing slopes, hollyleaf red berry (*Rhamnus ilicifolia*), orange monkey-flower (*Diplicus aurantiacus*), and Cedars creambush (*Holidiscus* sp.) are common. Also, coast silktassel (*Garrya elliptica*) is found in chaparral and serpentine reedgrass (*Calamagrostis ophitidis*) is an understory species. Some of the deeper soils among the cypress groves and chaparral support Douglas-fir and madrone (*Arbutus menziesii*) trees. Serpentine phacelia (*P. corymbosa*) is the most common plant in the more barren areas (Raiche 2009). White alder (*A. rhombifolia*) and ash (*Fraxinus latifolia*) are found in riparian areas and there are small ephemerally flooded areas with Brewer's willow (*Salix breweri*), azalea, hoary coffeeberry (*F.. californica* var. *tomentella*), and spicebush (*Calycanthus occidentalis*). Cottongrass (*Calliscirpus criniger*), California lady's slipper (*Cypripedium californicum*) and common maidenhair fern (*Adiantum capillis-veneris*) are found on alkaline seeps (Raiche 2009).

Mt. Tamalpais and Ring Mountain. Looking north from San Francisco, Mt. Tamalpais dominates the skyline, but at 734 m it is far from being one of the taller mountains in the California Coast Ranges. A small area of serpentinite is exposed on the crest of Mount Tamalpais, in the Mt. Tamalpais State Park. The serpentine plant communities are leather oak and mixed leather oak–Eastwood manzanita chaparral, and there are small stands of Sargent cypress.

Ring Mountain, east of Mt. Tamalpais, is an easily accessible ultramafic area on the Tiburon Peninsula in San Francisco Bay. With a summit at 183 m asl, Ring Mountain is more of a hill than a mountain. The slopes are mostly moderately steep to steep. Summers are dry and the mean annual precipitation, practically all rain, is about 750 mm. The ultramafic soils are on serpentinized peridotite and mélange with an ultramafic matrix (Rice and Wagner 1991). Huge blocks of blueschists and amphibolite of the Franciscan complex, weighing hundreds of tons each, jut above the ultramafic landscape (Fig. 10-7). Only Mollisols, mostly shallow soils (L. Argixerolls), with minor inclusions of other ultramafic soils, were mapped on Ring Mountain in the Marin County soil survey (Kashiwagi 1985).

Figure 10-7. Turtle Rock on Ring Mountain, a block of blue schist from mélange of the Franciscan complex.

The serpentine vegetative cover on Ring Mountain is grassland (Fiedler and Leidy 1987). Foothills and purple needlegrass *(Nassella lepida* and *N. pulchra*) and serpentine reedgrass (*Calamagrostis ophitidis*) are common grasses, and there are a few scattered clumps of California bay and coast live oak trees around rocky knobs, and some small patches of coyote bush

(*Baccharis pilularis* var. *consanguinea*) and leather oak. There are several rare endemic plants on Ring Mountain, and the Tiburon mariposa lily (*Calochortus tiburonensis*) is a major attraction for botanists. A native grassland on Ring Mountain has been set aside as part of the Nature Conservancy's California Critical Areas Program (Rice and Wagner 1991).

San Francisco. A strip of ophiolitic terrane nearly a kilometer wide runs from Hunters Point on San Francisco Bay across Potrero Hill to Baker Beach and the Golden Gate (Wahrhaftig 1984). The south end of the Golden Gate Bridge rests on serpentine mélange. The ultramafics have been urbanized, except in the Presidio, or the Golden Gate National Park (GGNP), and much of it has been covered by aeolian sand,. The mean annual precipitation, from south to north, ranges about 500 mm. All GGNP areas inland from the steep slopes above Baker Beach have been disturbed for recreational use or support groves of Monterey cypress (*Hesperocyparis macrocarpa*) and Monterey pine (*P. radiata*) that were planted there many years ago.

Soils on gentle to moderate slopes of the Presidio are cool (isomesic STR) Entisols, Inceptisols, and Mollisols that have moisture deficits in plant root zones for at least three consecutive months each year. Most of the soils have large amounts of aeolian sand in them and may not be ultramafic soils. Ultramafic soils on the steep slopes above Baker Beach are occupied by a dense cover of chaparral dominated by blue blossom (*Ceanothus thyrsiflorus*) and seaside wooly sunflower (*Eriophyllum staechadifolium*) and by grassland.

Jasper Ridge. Thin sheets of serpentine mélange are thrust across the Franciscan complex at the Jasper Ridge Biological Reserve (Coleman 2004, Page and Tabor 1967), which is just east of the San Andreas fault. The ultramafic landscape on Jasper Ridge is on about 1 km^2 of warm, summer-dry Mollisols (L. Haploxerolls and L. Argixerolls) on a low, gently rounded ridge. The mean annual precipitation is about 600 mm. The ultramafic Mollisols are mostly in grassland, and some are in leather oak–chamise chaparral around the margins of the grassland. Stanford University personnel and collaborators have done considerable research on the plants and animals of Jasper Ridge (for example, Hobbs and Mooney 1995, Dukes 2001, Ehrlich and Hanski 2004).

Coyote Ridge. Coyote Ridge is adjacent to the Hayward fault, at the south end of the fault zone, east of the Santa Clara Valley (Page et al. 1999). Serpentinite is exposed on the steep sides and gently sloping summits of Coyote Ridge. The summits are about 400 m asl and the mean annual precipitation is about 500 mm.

The soils in the grassland on Coyote Ridge are warm Mollisols, mostly lithic Haploxerolls of the Montara series and associated Argixerolls. Some shallow Mollisols, particularly lithic argillic Mollisols of the Henneke series, occur in chamise chaparral on steep sideslopes. Smectites are the dominant clay minerals in Mollisols sampled on Coyote Ridge (data of R Southard in a guidebook for the Western Society of Soil Science field trip, June 23, 1999). The serpentine grassland on Coyote Ridge is a core habitat for the Bay checkerspot butterfly (Harrison et al. 1988).

The vegetation of Coyote Ridge has been thoroughly studied by McCarten (1992) and Evens and San (2004). It includes extensive grassland and herbaceous communities dominated by both native and nonnative species. Despite the common resistance of ultramafic areas to invasion by exotic plant species elsewhere, the Coyote Ridge locality shows significant incursion by nonnative species. Apparently native herbaceous stands have been recently type converted to nonnative grasslands, with Italian ryegrass (*Lolium multiflorum*) as one of the dominant species. Air pollution and nitrogen deposition from the adjacent heavily urbanized Santa Clara Valley have contributed significantly to this conversion (Weiss 1999). Nevertheless, extensive areas of largely native herbaceous vegetation still exist at Coyote Ridge. Some of the characteristic native herbaceous species are the serpentine indifferent dwarf plantain (*Plantago erecta*), small fescue (*Vulpia microstachys*), goldfields (*Lasthenia californica*), erect dwarf-cudweed (*Hesperevax sparsiflora*), and hayfield tarweed (*Hemizonia congesta* ssp. *luzulifolia*). There are also wetland and seep communities dominated by native herbs; one of the most distinctive being the endemic Mt. Hamilton thistle (*Cirsium fontinale* var. campylon), a rare and endangered species. There is some scrub and chaparral on ridge sideslopes, where leather oak (*Q. durata*) is dominant, along with other serpentine chaparral species. Bigberry manzanita (*Arctostaphylos glauca*) occurs on the slopes with a variety of other shrubs, including coastal sage scrub species such as California sagebrush (*Artemisia californica*), black sage (*Salvia mellifera*), and coyote ceanothus (*Ceanothus ferrisiae*).

10B4 Coastal belt

The Coastal belt lacks ultramafic rocks. It may have been emplaced too late to have received ultramafic rocks overthrust from the Coast Range ophiolite

10B5 Diablo Range

Soils and plants of the Mt. Diablo and New Idria serpentinites are representative of the kinds of them in the Diablo Range. The ultramafic rocks on Mt. Diablo were thrust from the Coast Range ophiolite over the Franciscan mélange, and the Clear Creek (or New Idria) serpentine is in a diaper that was pushed up from the Franciscan complex below through overlying exposures of Cenozoic rocks during the Neogene. Barrens are extensive on the New Idria body. Mean annual precipitation ranges mostly from 400 to 700 mm, but down to the 200-250 mm range in the Big Blue Hills.

Mt. Diablo. The ultramafic rocks on Mt. Diablo are in the Coast Range ophiolite, where it is separated from rocks of the Franciscan complex by a fault (Unruh et al. 2007). Serpentinite, with minor amounts of serpentinized peridotite and pyroxenite, is exposed on a narrow SW-NE strip about 2 km north of the 1174 m high summit of Mt. Diablo (Pampeyan 1963).. The ultramafic exposure is about 0.5 km wide and 8 km long; it separates mainly diabase and basalt of the Coast Range ophiolite on the northwest from rocks of the Franciscan complex on the southeast. Altitudes on the serpentinized peridotite range from about 210 to 760 m. Slopes are mostly steep, but locally gentle, particularly on Long Ridge.

The ultramafic soils are predominantly warm, lithic Mollisols with argillic horizons (Lithic Argixerolls); mainly soils of the Henneke series. Its distribution is shown in Figure 6 of Ertter and Bowerman (2002). The Hennecke soils occur in both grassland and chaparral on Mt. Diablo. Because blue oak avoids serpentine, the only trees are sparse gray pine. The main shrubs are whiteleaf manzanita (*A. viscida*), chamise (*A. fasciculatum*), toyon (*Heteromeles arbutifolia*), leather oak (*Q. durata*), Jim brush (*Ceanothus oliganthus* var *sorediatus*), and buckbrush (C. cuneatus). The predominant grass is wild oat (*Avena* spp.), with common to sparse bluegrass (*Poa secunda*), brome (*Bromus* spp.), California fescue (*F. californica*), and onion grass (*Melica torreyana*) in grassland and among the shrubs, and with goldback fern (*Pentagramma triagularis*) among the rocks.

New Idria. New Idria is the name of a mercury mining area that is marginal to a large ultramafic body. The mercury mining ceased in 1970, because of water pollution downstream from the area. Highly sheared ultramafic rocks are exposed in an area about 20 km long and 5 km wide in an anticlinal form along the northeast side of the San Andreas fault (Fig. 10-1A, Coleman 1996). The serpentinite is in a feature that has been described as a diapir that was pushed up from the Franciscan complex through post-Franciscan rocks of the Diablo Range during the Neogene. The serpentinite is completely serpentinized peridotite. It is mostly chrysotile with some lizardite, brucite, and magnetite. Most of the serpentinite is in the Clear Creek Management Area of the Bureau of Land Management. Public access to the CC Management Area has been limited for several years, because the asbestos there is perceived to be a great health hazard.

Altitudes on the New Idria serpentinite are about 800 m near Clear Creek to 1602 m on San Benito Mountain, which is the highest elevation in the Diablo Range. The mean annual precipitation is in the 400 to 600 mm range and the summers are hot and dry. CC Management Area soils that are in Fresno County have been mapped (Arroues 2006), and more recently Kerry Arroues mapped the CC Management Area soils in San Benito County (personal communication, Arroues 2014). Ultramafic soils in large parts of the area are very shallow Entisols that lack vegetation. The ultramafic barren areas are the largest in North America. Warm soils at the lower elevations are mostly shallow Mollisols (Lithic and shallow Typic Argixerolls), and there are some deeper Alfisols. Cool soils at the higher elevations are mostly shallow Alfisols (Mollic Haploxeralfs) and deeper Mollisols (Argixerolls). Griffin (1975) found that the most common shrubs were Baker's manzanita (*A. bakeri*), buckbrush (*C.. cuneatus*), leather oak (*Q. durata*), rubber rabbitbrush (*Ericameria nauseosus*), and hollyleaf redberry (*Rhamnus illicifolia*). Other plants common at the lower elevations were chamise (*A. fasciculatum*), scrub oak (*Quercus* sp.), bigberry manzanita (*A. glauca*), birchleaf mountain mahogany (*C. betuloides*), toyon (*H. arbutifolia*), Spanish bayonet (*Yucca whipplei*), and small fescues (*Vulpia* spp.). Gray pine trees occur sparsely at the lower elevations and Coulter pine (or a hybrid of Coulter and Jeffrey pines) and incense cedar are common at the higher elevations.

Rajakaruna et al. (2012) found 119 species of lichens and fungi on three different kinds of rocks in the New Idria area. Serpentinite, with a total of 60 species, had much more diverse lichen communities than silica-carbonate or sandstone and shale. Only four species occurred on all kinds of these rocks.

Big Blue Hills. About 5 km^2 of serpentinite debris that slid off of the New Idria body during the Neogene is exposed in an olistrostome of the Big Blue Hills (Fig 10-1A). The Big Blue Hills are on the southwest margin of the San Joaquin Valley, about 15 km east and 1200±150 m lower than San Benito Mountain, which is on the New Idria serpentine. Summer temperatures are hot in the Big Blue Hills and the mean annual precipitation is in the 200 to 250 mm range.

The Big Blue debris deposits consist of serpentinite debris-flows that grade into serpentinite-clast conglomerate and breccia that in turn grade into serpentine-flake mudstones and laminated serpentinite grain sands (Dickinson and Casey 1976). It is estimated that the initial erosion of the New Idria serpentinite continued through middle to late Miocene, depositing more than 5 cubic miles (a volume >20 km^3) of serpentinite debris.

The dominant soil on the weakly consolidated Big Blue deposits is a shallow argillic Mollisol with an A/Bt/Cr horizon sequence (Arroues 2006). Its cover is annual grassland, with common shrubs. The shrubs are mostly allscale (*Atriplex polycarpa*) and sparse California buckwheat (*Eriogonum fasciculatum*). Rattail fescue (*Vulpia myuros*) is the dominant grass, with common soft chess (*Bromus hordeaceus* ssp. *hordeaceus*) and mouse barley (*Hordeum murinum*) and sparse red brome (*B. rubens*) (Natural Resources Conservation Service, NRCS). The estimated annual forage production was 900±300 pounds/acre (1020 kg/ha).

Table Mountain. Table Mountain is a ridge several km long, with a rounded summit that is a few hundred meters wide. Its summit, at about 1000 m asl, is 100 to 200 m below, and parallel to, the crest of the main ridge that runs southeast-northwest along the summit of the Diablo Range. Parkfield, 5 km southwest of Table Mountain and about 450 m lower, is on the San Andreas fault where it runs along the bottom of the Cholame Valley. A serpentine mélange with serpentinized peridotite clasts was extruded from the Franciscan complex and spread across Table Mountain during the Neogene (Dickinson 1966). Subsequently, mass wasting and fluvial erosion have carried serpentinite downslope, toward Turkey Flat, which is on the northeast side of the Cholame Valley. Elevations of the serpentinite along Table Mountain are about 900 to 1000 m, and elevations of ultramafic soils on the sides of Table Mountain range down to about 450 m. The mean annual precipitation is about 400 to 500 mm.

The main soil on the gentle slopes of Table Mountain is Montara (Cook 1978). It is a shallow or lithic haplic Mollisol (Haploxeroll), and the plant cover is grassland. Other ultramafic soils mapped on steeper slopes are shallow or lithic argillic Mollisols (Argixerolls) and Vertisols. The serpentine mélange, or colluvium from it, occurs along the Parkfield-Coalinga road, a short distance northwest of Table Mountain. The serpentinite there is so closely scrambled with other Franciscan rocks, that it is difficult to decide which soils and plant communities are predominantly ultramafic. There is much chamise and leatheroak, and grassland. The chamise appears to be present only with some nonserpentine influence. There is sparse gray pine, and some western juniper (*J. occidentalis*), which appears to tolerate more serpentine influence than the blue oak (*Q. douglasii*).

An NRCS range site was inventoried on the Climara soil, an ultramafic Vertisol (Arroues 2006). The cover was annual grassland. mostly wild oat (*Avena fatua*) and soft chess (*Bromus hordeaceus* ssp. *hordeaceus*) with rattail fescue (*Vulpia myuros*), red brome (*B. rubens*), sparse pine grass (*Poa secunda*), and purple needlegrass (*Nassella pulchra*). The estimated annual forage production was 2800±900 pounds/acre (3185 kg/ha), which is quite high compared to the more definitely ultramafic soil on Big Blue of the Blue Hills..

10B6 Nacimiento Sector

The Nacimiento sector has been separated from the Central Belt and Diablo Range, which have similar terranes, by northward movement of the Salinian block between the San Andreas fault and the Sur-Nacimento fault zone (Fig. 10-3). There are both ultramafic bodies that are rooted in the Franciscan complex (for example Burrow Mountain, Burch 1968) and ultramafic bodies that have been thrust from the Coast Range ophiolite over the Franciscan complex (for example, on Cuesta Ridge, Snow 2002). All of the ultramafic soils are warm and summer-dry, and the vegetation is mostly chaparral and grassland.

Cuesta Ridge. Highly serpentinized ultramafic rocks are exposed for about 25 km along the northwest trend of Cuesta Ridge (Fig. 10-1A), east of Moro Bay, and on steep slopes down from the narrow ridge. The ultramafic rocks are dismembered oceanic crustal components of Coast Range ophiolite that were thrust over the Franciscan complex and Monterey formation during the Neogene, before Miocene sediments of the Monterey formation were lithified well (Snow 2002).

Altitudes on the ultramafic rocks are up to about 800 m on Cuesta Ridge. The mean annual precipitation is about 500 to 600 mm. Ultramafic soils along Cuesta Ridge are shallow, or lithic, and moderately deep Mollisols (Lithic and Typic Argixerolls) with chaparral and patches of Sargent cypress (O'Hare and Hallock 1980). There are some ultramafic Vertisols, and grassland, down the southwest side of Cuesta Ridge toward San Luis Obispo. Many ultramafic soils on the southwest side of Cuesta Ridge are deep, possibly in ancient colluvial deposits.

Figure 10-8. The Cuesta Ridge Botanical Area.

The Cuesta Ridge Botanical Area (Fig. 10-8) was established on the ridge to preserve an example of the Sargent cypress (*Hesperocyparis sargentii*) woodland and associated rare serpentine plants such as Bishop manzanita (*A. obispoensis*). Frequent burning favors Sargent cypress, which has serotinus cones. Among the local endemics on serpentine are San Luis Obispo lily (*Calochortus obispoensis*), Cuesta Pass checkerbloom (*Sidalcea hickmanii* ssp. *anomala*), and San Luis Obispo sedge (*Carex obispoensis*). Common shrubs in the serpentine chaparral are leatheroak (*Q. durata*), buckbrush (*C. cuneatus*), toyon (*Heteromeles arbutifolia*), and bush poppy (*Dendromecon rigida*); chamise is a minor component of the serpentine chaparral. These shrubs on Cuesta Ridge are the ones that Wells (1962) found to be most common on other ultramafic soils in San Luis Obispo County, and he found California bay (*Umbellularia californica*) to be a common tree, and at lower elevations, coast live oak (*Q. agrifolia*). The foothill needlegrass (*Nassella lepida*) was dominant in serpentine grassland (Wells 1962).

Figueroa Mountain and the Sedgewick Reserve. Figueroa Mountain is on the southwestern side of the southeast-northwest oriented San Rafael Mountains, 24 km northeast of Buellton. The summit of Figueroa Mountain is on Miocene sedimentary rocks. The ultramafic rocks are in mélange of the Franciscan complex, but there are some rocks from the Coast Range ophiolite and Great Valley sequence north of Figueroa Mountain (Vedder et al. 1980). The serpentinized peridotite in the Franciscan mélange is intermingled with metagraywacke, metaconglomerate, metachert, greenstone, glaucophane schist, and amphibolite; it is in a strip about 2 to 5 km wide that reaches 30 or 40 km southeast from Figueroa Mountain nearly to the Santa Ynez Mountains. The ultramafic rocks and plant communities between about 600 and 1200 m asl southwest of Figueroa Mountain, on the flank of the San Rafael Mountains, are considered here.

Ultramafic exposures on the southwest side of Figueroa Mountain are steep. The mean annual precipitation is in the 400 to 500 mm range, and the summers are hot and dry. Most of the soils are shallow or lithic Mollisols and deeper Vertisols, and there are some Entisols. It seems as though every exposure of the ultramafic rocks and soils has a different plant community. They range from

Spanish bayonet (*Yucca whipplii*) on extremely rocky slopes, to sparsely vegetated, very shallow soils with bristleleaf goldenbush (*Hazardia stenolepis*), bluegrass (*Poa secunda*), and annual fescue (*V. microstachys*), to shallow soils with dense stands of bigberry manzanita (*A. glauca*), and other shrubs, and deeper soils with chaparral containing buckbrush (*C. cuneatus*), toyon (*H. arbutifolia*), holly-leaf redberry (*Rhamnus ilicifolia*), and sparse gray pine, and to Vertisols with grassland dominated by wild oat (*Avena* spp.), soft chess (*B. hordeaceus*), and red brome (*B. rubens*).

The upper part of the 2388 ha (967 acre) University of California's Sedgewick Reserve, which is a few kilometers southwest of Figueroa Mountain, is at the edge of the San Rafael Mountains and the lower part of te Reserve is in foothills, down from the mountains. There might be some moderately deep ultramafic Vertisols on the upper margin of the Sedgewick Reserve, but no other ultramafic soils were mapped in the Reserve area by the NRCS.

The Figueroa Mountains area is well known for reliable and often spectacular spring wildflower displays (Smith 1998) and for endemic serpentine species. The vegetation includes grassland with a high native component and chaparral, including the southernmost stands of leather oak. Santa Barbara jewelflower (*Caulanthus amplexicaulis* var. *amplexicaulis*) is a notable serpentine endemic in this region. Publications on the serpentine vegetation and flora of the Sedgewick Reserve include those of Gram et al. (2004) and Seabloom et al. (2003).

C Baja California

Mesozoic to early Cenozoic accreted terranes south from the Santa Cruz and Santa Rosa Island faults past Santa Catalina Island and along the Pacific coast of Baja California (Fig. 10-1B) are much like those in the California Coast Ranges. They are present on the Vizcaíno Peninsula and on San Benito, Cedros, Magdalena, and Margarita Islands (Sedlock 2003). Santa Catalina Island, which is south of the Santa Barbara Channel but still called a Channel Island, has exposures of rocks resembling those of the Franciscan complex (Norris and Webb 1990). The accreted terrane of Calmalli (Fig.10-1B), which is about midway between Guerrero Negro and the Gulf of California, contains ultramafic rocks (Alexander et al. 2007a), but it has a different geologic origin and is considered in Chapter 11 (Mexico), rather than with the accreted terranes on the Pacific coastal side of Baja California.

The climates of Pacific coastal Baja California are warm temperate to hot subtropical, with moderate January to July air temperature differences of 6.6°C (12.5-19.1°) at Avalon on Santa Catalina Island to 5.6°C (18.6-24.2°) on Cedros Island, and 5.7°C (17.0-23.7°) at San Carlos on Magdalena Bay. Climates are summer-dry on the Channel Islands, arid with little summer rain on Cedros Island and the Vizcaíno Peninsula, and arid with appreciable late summer rain on Magdalena and Margarita Islands.

10C1 Santa Catalina Island

Oceanic terranes are believed to exist beneath all of the Pacific Islands south of the Santa Cruz fault (Norris and Webb 1990), but ultramafic rocks are exposed only on Santa Catalina Island and islands of Baja California. There are small amounts of serpentinite and soapstone in Catalina schist exposures on the northern half of 194 km^2 Catalina Island. They are components of oceanic crust that dates from the Triassic to Jurassic and were accreted beginning in the Cretaceous Period. Santa Catalina aborigines excavated soapstone and serpentine to make utensils and ornaments (Wlodarski 1979).

The mean annual rainfall ranges from <300 mm at lower elevations on the east side of Santa Catalina Island to nearly 450 mm at higher altitudes, which are up to 639 m asl. Soils on the island are warm (thermic STRs) and summer-dry. No ultramafic soils were mapped on Santa Catalina Island (Ballmer 2008), because the areas (individual polygons) of them were too small (Ballmer 2014, personal communication). There are many plant species that are unique on Santa Catalina Island, but no serpentine was mentioned in a flora of the island (Thorne 1967).

10C2 Vizcaíno Peninsula

The greatest concentration of the ultramafic rocks in Baja California is in the mountainous terrain from Cedros Island southward along the coastal side of the Vizcaíno Peninsula nearly to Bahia Asunción. These ultramafic rocks are associated with island arc and seafloor deposits that were accreted to the North American continent during the Mesozoic era (Dickinson and Lawton 2001, Sedlock. 2003). The Vizcaíno Peninsula is in the Sonoran Desert, as it was described by Shreve (1951) and Wiggins (1980).

Summer temperatures are warm to hot and winter temperatures are cool. The soil temperature regimes are marginal between hyperthermic and isohyperthermic STRs. The mean annual precipitation is about 100 mm from Cedros Island southward along the coastal side of The Vizcaíno Peninsula (Hastings and Turner 1965). Drought persists for most of each year in the ultramafic areas (Hastings and Humphrey 1969).

Shallow Entisols (Lithic Torriorthents), and less extensive deeper Entisols in colluvium, dominate the arid mountain slopes of ultramafic rocks on the Peninsula, with some weakly developed Aridisols (Haplargids) on moderately steep slopes and on colluvial footslopes (Alexander 2007). Argillic horizons are the only evident diagnostic ones on the mountain slopes, other than ochric epipedons, but duripans are present in piedmont and terrace soils. Some of the ultramafic Aridisols are calcareous, and some on footslopes have Btk horizons, but none are known to have sufficient carbonate accumulations for calcic horizons. Chunks of magnesite that are present in many of the Aridisols are fragments from veins in mafic or ultramafic bedrock. Magnesite has been mined in several places where the veins are large enough for miners to collect large blocks of it for export (Fig. 2-7). Some of the alluvium in washes, or arroyos, in the ophiolite of Calmalli, and on the Vizcaíno Peninsula, is cemented with silica.

Vegetation is sparse in the ultramafic habitats. It is mostly shrubs, few trees, cacti, and sparse yucca. The lone tree is copalquin (*Pachycormus discolor*). There are many kinds of shrubs, palo Adán (*Fouquieria diguetii*), several species of cacti, and agave (*Agave* spp.). Forbs are common in the spring, following winter rain, and in the early autumn following late summer rain. Some of the more common shrubs are Pacific errazurizia (*Errazurizia benthamii*), buckwheats (*Eriogonum* spp.), spurge (*Euphorbia magdaleneae*), lentisco (*Rhus lentii*), snapdragon (*Gambelia juncea*), jojoba (*Simmondsia chinensis*), and cañatillo (*Ephedra* sp.). Cenizo (*Atriplex canescens*) is common on some ultramafic footslopes. Pitaya agria *(Stenocereus gummosus*) is one of the more common cacti on the ultramafic soils.

D An Arid Terrace Sequence

Streams that are cutting downward commonly cut down episodically, leaving older deposits on a sequence of terraces above fluvially active stream channels or floodplains.

To ascertain the course of ultramafic soil and landscape development in the arid climate of Baja California, soils and plant communities were examined on a sequence of fluvial terraces of a piedmont that slopes northeastward from low mountains at the northern end of the Sierra San Jose de Castro (Fig. 10-9A). The piedmont represents a very old fluvial plain that has been entirely dissected by water from streams that flowed from the low mountains. Some of the streams cutting though the piedmont had floodplains that became terraces when the streams cut down farther. The soils and plant communities were described at four sites from an ephemerally active wash (site 1) to sites on successively higher terraces (sites 2, 3, and 4; Fig. 10-9). Investigation of the soils and plant communities in this sequence indicates how the arid soils and plant communities might have evolved through the past. Although the soils were not dated, their development is expected to have spanned hundreds of thousands of years.

10D1 The alluvial fans

Sites 1, 2, and 3 were examined in alluvium at about 350 m asl, 0.3 to 0.4 km from the San José de Castro mountain front, on surfaces that slope northeast with 5 to 6% gradients (Fig 10-10B). Nearly a kilometer southeast of these three sites, a fourth site was examined on an old terrace that is inset 6 to 8 m below the ridges of a very old, dissected piedmont plain (Fig 10-10A). The plain represented by the ridges is about 12 to 15 m higher than sites 1, 2, and 3. These ridges are the remnants of very old alluvial fans, or a pediment, that has been completely dissected, removing the older surface that might have had very old soils on it. Gullies in the very old fan deposits expose gravels cemented by silica to depths >8 m. The surface of the inset fan with site 4 on it is being dissected, but it still has a flat surface with old soils.

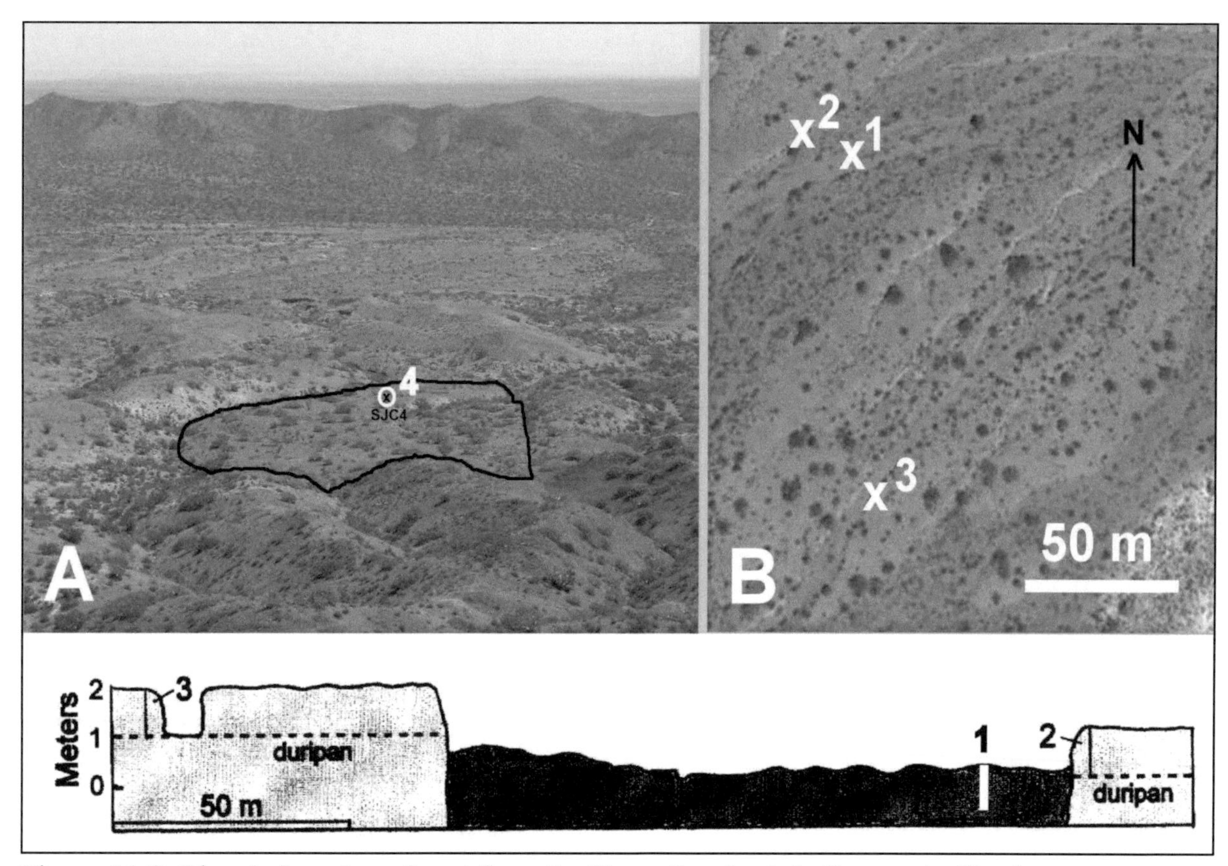

Figure 10-8. The piedmont northeast from the Sierra San José de Castro. A. Site 4 on a terrace inset below the dissected piedmont plain. B. Sites 1, 2, and 3 on a floodplain and terraces. Below. A cross-section through the floodplain and terraces of sites 1, 2, and 3.

Site 1 is in an ephemeral wash that has an undulating surface due to numerous shallow runoff furrows (Fig 10-9). The alluvial fan terraces have planar surfaces that are dissected by few drainage channels. All of the larger drainage channels commence on the mountain upslope from the alluvial fans of the piedmont. The channel that crosses the terrace at pedon 3 is up to 10 m wide and it is about 1 m deep at pedon site 3, where runoff has cut down to a duripan.

All of the alluvial fan deposits are very gravelly and slightly cobbly, with subangular to subround pebbles and cobbles. Cobbles cover 1 to 3% of the terrace surfaces. Pebbles and cobbles in the fluvial parent materials of the soils are about 90% serpentinized peridotite, with traces of gabbro, graywacke (dirty sandstone), and magnesite. On the ground, pebbles and cobbles exposed to the atmosphere are mostly reddish brown (2.5YR 5/5 dry) on the site 4 alluvial fan, patchy or partially reddish brown on the site 3 fan, sparsely reddish brown on the site 2 fan, and pebbles and cobbles of the wash lack reddish brown weathering surfaces. There is no desert pavement. Relative soil ages correspond sequentially to the elevations of the surfaces they are on, although it is not certain that the pedon 4 soil is older that the pedon 3 soil. Pedon 4 is about 4 m higher than pedon 3, but it is along a different stream channel that has had a different geomorphic history.

Soils were described with the Soil Survey Manual (Soil Survey Staff 1993) as a guide. Plants were identified by reference to Wiggins (1980), with some more current taxonomic names from Rebman and Roberts (2012). Current plant names are those listed by Rebman et al. 2016

10D2 The soils and vegetation

Soil parent materials. Sand was separated from the terrace soils and examined with a polarizing microscope. The coarser sands were mostly subangular to subround, gray rock fragments, with many fragments weathered yellowish- to reddish-brown in the subsoils on the terraces. Very fine sand grains were separated with bromoform (specific gravity, SG=2.89). About five to ten percent of the very fine sand fractions were grains heavier than bromoform, including some that were magnetic and presumed to be magnetite. The few heavy grains, other than magnetite, that were not too weathered to be identifiable were mostly clinopyroxene and some hornblende. More than 90% of the grains floated on bromoform. The light grains were mostly serpentine, including some chlorite, but with substantial quartz and some feldspars. Because the pebbles and coarser sand grains in the terrace soils lacked quartz, quartz grains in the very fine sands were assumed to be from aeolian sources.

Figure 10-10. Terrace soils and a common plant. A. Pedon 2 on the lowest terrace, with a basalt hill beyond the terrace. B. Pedon 4 on the highest terrace . C. Rough sunflower (*Encelia asperifolia*) on the low terrace, near Pedon 2.

Table 10.1 Soil properties from pedon descriptions, and pH for fine-earth (soil < 2 mm).

Horizon		Munsell Color		Texture (feel)	Structure	pH
sym.	depth	moist	dry			DW
Pedon 1, Torrifluvent						
A	0-3 cm	10YR 3/3	5/4	extremely gravelly sandy loam	fine granular	7.2
AC1	3-18	10YR 3/3	6/4	very gravelly sandy loam	massive	7.4
AC2	18-54	10YR 3/3	6/4	very gravelly loamy sand	massive	7.6
C	54-110	10YR 4/3	6/4	very gravelly loamy sand	massive	7.7
Pedon 2, Typic Argidurid						
A1	0-1 cm	7.5YR 3/4	5/4	very gravelly fine sandy loam	medium platy	7.2
A2	1-14	7.5YR 3/4	5/5	gravelly loam	massive	7.5
Bt [a]	14-27	7.5YR 4/4	5/5	gravelly clay loam	subang. blocky	7.7
BC	27-68	7.5YR 4/3	6/4	very gravelly sandy loam	massive	8.2
C	68-102	10YR 4/3	6/3	very gravelly loamy sand	massive	8.4
Cqm	102-120	10YR 4/3	6/3	—	—	—
Pedon 3, Typic Argidurid						
A1	0-3 cm	5YR 4/6	5/6	gravelly sandy loam	coarse platy	7.3
A2	3-8	5YR 4/6	5/6	gravelly sandy loam	subang. blocky	7.4
Bt [a]	8-21	5YR 4/6	5/6	gravelly sandy clay loam	subang. blocky	7.5
Btq[b]	21-45	7.5YR 4/4	6/4	very grav. sandy clay loam	subang. blocky	7.9
BCq	45-70	10YR 4/3	6/4	very gravelly sandy loam	massive	7.8
Cq	70-110	10YR 4/2	6/3	very gravelly loamy sand	massive	7.4
Cqm	110-120	10YR	7/2	—	—	—
Pedon 4, Duric Petroargid						
A1	0-3	7.5YR 3/4	5/5	very gravelly loam	fine granular	7.6
A2	3-14	7.5YR 4/4	5/5	very gravelly loam	subang. blocky	7.5
Bt1	14-26	5YR 4/5	5/6	very gravelly clay loam	subang. blocky	7.6
Bt2	26-45	7.5YR 4/4	5/5	very gravelly sandy clay	subang. blocky	7.6
BCq	45-88	10YR 4/3	5/3	extr. grav. sandy clay loam	massive	7.0
Cq	88-105+	10YR 4/2	6/2	extremely grav. sandy loam	massive	—
Cqm	below 105 cm and above 150 cm, observed in a gully cut into surface 4					

[a] Many thin ped coatings 2.5YR 4/6, 3/6 moist. [b] Common thin ped coatings 2.5YR 4/6, 3/6.

Large amounts of aeolian sand input were inferred from the proportions of very fine sand, which increase from the floodplain to the surface horizons of the terrace soils. For a more definitive resolution of this assumption, 300 grain samples of light (SG<2.89) very fine sand were identified and tallied. About one-third of the grains were quartz in the surface horizons of all soils, but down nearly to 20% in the terrace subsoils (about 25-45 cm depths, Alexander 2019).

Soil properties. Weathering and leaching of the terrace soils has produced neutral to moderately alkaline soils with subsoil clay and silica accumulations (Table 10.1). The main features of soil development have been the accumulation of clay in reddish argillic (Bt) horizons above 0.5 meter depths (Fig. 10-10) and silica cementation in hardpans (duripans, Cqm) below about one meter. None of the soils were calcareous. The most prominent clay minerals were serpentine in the floodplain soil and palygorskite in the terrace soils (Fig III-5).

Table 10.2. Chemical analyses of the soils.

Pedon and Horiz.	Soil Depth	Exchange. cations			Aqua Regia digest									
		Ca	Mg	Ca/Mg	g/kg				Ca/Mg	g/kg		mg/kg		
	cm	mmol+/kg		molar	Na	K	Ca	Mg	mass	Al	Fe	Cr	Co	Ni
1-1	0-3	73	42	1.74	0.3	1.2	3	62	0.05	7	38	304	69	971
1-2	3-18	72	42	1.71	0.3	1.1	3	60	0.05	8	39	298	65	926
1-3	18-54	65	48	1.35	0.5	0.6	5	91	0.05	9	52	433	74	1312
1-4	54-110	69	54	1.28	0.3	0.9	5	70	0.07	7	40	339	66	1041
2-1	0-1	51	76	0.67	0.2	1.9	2	33	0.06	11	46	326	50	955
2-2	1-14	56	81	0.69	0.2	2.0	2	32	0.06	11	48	340	49	994
2-3	14-27	71	76	0.93	0.2	1.1	2	45	0.04	12	59	477	45	1332
2-4	27-68	79	64	1.23	0.3	0.7	3	40	0.08	8	35	301	30	820
2-5	68-102	76	56	1.36	0.7	0.4	2	65	0.03	8	41	353	37	1073
3-1	0-3	52	82	0.63	0.1	1.0	1	21	0.05	7	34	266	35	798
3-2	3-8	64	103	0.62	0.1	1.3	1	17	0.06	8	36	276	36	840
3-3	8-21	80	104	0.77	0.1	1.0	1	16	0.06	8	40	340	35	973
3-4	21-45	87	68	1.28	0.6	0.6	2	27	0.07	7	41	377	33	969
3-5	45-70	73	62	1.18	0.9	0.4	1	43	0.02	6	35	340	35	982
3-6	70-110	74	64	1.16	1.1	0.3	5	62	0.08	5	37	363	45	1017

Note. Soil samples with Ca contents of 50 to 100 mmol+/kg would have Ca concentration of 1 to 2 g/kg of soil, which is near the lower limit of detection for the aqua regia results. Thus, the terrace soils may have practically negligible amounts of Ca other than exchangeable Ca.

Analyses from aqua regia digestion show that, other than Si and oxygen (not quantified), Fe and Mg are the main chemical elements in the soils (Table 10.2). Silicon has been leached downward to produce duripans, and Si is a major constituent of the clay minerals that have accumulated in the subsoils. Iron oxides (mainly goethite, FeOOH) have accumulated and, with hematite, have imparted reddish hues to the subsoils. As usual in ultramafic soils, the concentration of Cr, Co, and Ni are very high and Al is low, compared to other soils (Shacklette and Boerngen 1984). There may have been scarcely any Ca in the serpentinized peridotite soil parent material, and most of that could have been in clinopyroxenes that were readily weathered, releasing the Ca to be leached from the soils. The exchangeable Ca might be replenished by enough Ca in aerosols or dust to maintain exchangeable amounts of Ca similar to Mg, but not sufficient to produce Ca-carbonates.

A few soil samples from pedons 1, 2, and 3 were sent to the A&L Agricultural Laboratories in Modesto, CA, where exchangeable cations were extracted with neutral, molar ammonium-acetate. The exchangeable K was 7 to 12 mmol+/kg in the surface and 2 to 9 mmol+/kg in the subsoils. Exchangeable Na was 2 to 3 mmol+/kg in the surface and 9 to 17 mmol+/kg in argillic horizons, and up to 44 mmol+/kg below 45 cm in pedon 3.

Vegetation. A striking feature of the vegetative cover is that there are no trees in the ephemeral wash (around site 1), even though copalquín (*Pachycormus discolor* var. *veatchiana*) occurs on the low terrace (around site 2) and is common on the older terraces (around sites 3 and 4). The plant species distributions are similar on the three terrace surfaces, but somewhat different on the mountain slopes and in the ephemeral wash (Table 10.3). Common saltbush (*Atriplex polycarpa*) in the wash and its absence on the mountain slopes is one of the more obvious differences. Rough sunflower (*Encelia asperifolia*) is the most common plant (Table 10.3), and dodder (*Cuscuta* sp.) is common on the plants. The recycling of nutrient elements in most ultramafic geoecosystems commonly adds enough Ca in plant detritus to soils that the surface horizons have higher exchangeable Ca/Mg ratios than subsoil horizons, but that trend is not evident in these sparsely vegetated San José de Castro ultramafic terrace soils (Table 10.3).

Table 10.3 Common plant species on low San José de Castro mountain slopes (upland), recent wash (site 1), and terraces of sites 2, 3, and 4. Visually estimated cover (percent by species).

Plant species	Upland	Site 1	Site 2	Site 3	Site 4
Tree - *Pachycormus discolor*	2	0	1	3	5
SS - *Fouquieria diguetii*	0	<1	2	2	1
S - *Ambrosia dumosa*	3	<1	<1	<1	3
S - *Atriplex polycarpa*	0	1	<1	<1	<1
S - *Bebbia juncea*	0	1	0	0	0
S - *Errazurizia benthamii*	<1	<1	0	0	0
S - *Euphorbia misera*	0	0	0	2	1
S - *Peritoma arborea*	1	1	0	<1	<1
S - *Lycium* sp.	<1	0	0	0	0
S - *Mirabilis laevis* var. *crassifolia*	2	2	1	<1	0
S - *Petalonyx linearis*	0	1	0	0	0
S - *Simmondsia chinensis*	<1	0	0	0	0
S - *Solanum hindsianum*	1	0	0	0	0
S - *Bahiopsis (parishii?)*	0	0	0	2	<1
S - *Encelia asperifolia*	3	8	4	2	2
C - *Lophocereus schottii*	0	<1	<1	<1	<1
C - *Ferocactus peninsulae* var. *vizcaínensis*	<1	0	0	<1	<1
F - *Chaenactis* sp.	0	1	<1	<1	<1
F - *Chorizanthe* sp.	<1	1	<1	<1	<1
F - *Cryptantha* sp.	3	4	2	<1	<1
F - *Cuscuta* sp.	<1	10	8	2	1
F - *Lasthenia* sp.	<1	<1	<1	<1	<1
F - *Lupinus* sp.	<1	<1	<1	<1	<1
F - *Sibara angelorum*	2	1	<1	<1	<1
F - *Phacelia* sp.	<1	0	0	0	0
F - *Plantago ovata*	<1	1	1	1	1
F - Portulacaceae	<1	1	<1	<1	<1
F - *Stephanomeria* sp.	<1	2	1	<1	<1

Abbreviations: C, cactus; F, forb; S, shrub; and SS, stem succulent.

11 Mexico

Mexico is a geologically, topographically, and climatically diverse country. It is dominated physiographically by a Cordillera Occidental east of the Gulf of California, a Cordillera Oriental above a narrow Gulf of Mexico coastal plain, an interior plateau between these north-south oriented mountain ranges and a Trans-Mexico Volcanic Belt (Cinturón Volcánico, or Eje Neovolcánico) south of the interior plateau, at about 20°N latitude (Fig. 11-1). The Sierra Madre del Sur, or Cordillera del Sur, along the south coast of Mexico is separated from the Trans-Mexico Volcanic Belt by a deep valley called the Depresión de Bolsas. Baja California, west of the Gulf of California has been separated from the rest of Mexico and translated northward (dextrally) along a Cenozoic fault system.

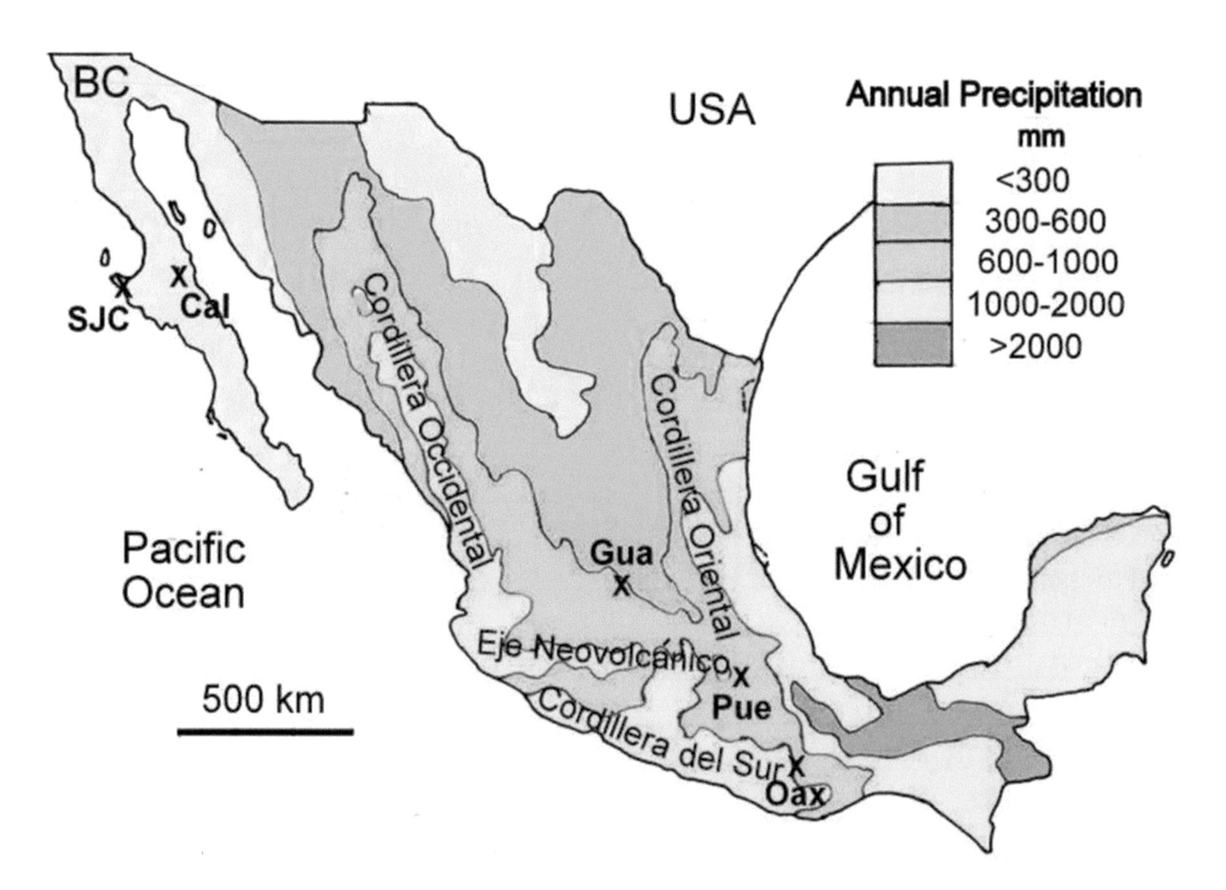

Figure 11-1. Major mountain ranges in Mexico and sites of soil and plant sampling. BC, Baja California; CAL, Calmalli; SJC, San José de Castro.

The geology of Mexico is very complex. It has evolved over a billion years (Keppie 2004). There are ultramafic bodies in many of the terranes in Mexico, but none of them are large, with the exception of the Vizcaíno terrane on the Vizcaíno Peninsula and Cedros Island where the ultramafic terrain covers several hundred km^2 (Chapter 10, Coastal Southwest, section 10D). The ultramafic bodies are widely distributed across Mexico west of the Cordillera Occidental and south of the Cinturón Volcánico (Ortiz-Hernandez et al. 2006).

Mexico has a great range of climates and vegetation, from hot deserts on the northwest to cooler deserts, high mountains, and subtropical forests. There are large areas that are above 2,000 meters in the Cordillera Oriental and Cordillera Occidental on the east and west, and in the Trans-Mexico Volcanic Belt and the Sierra Madre del Sur in southern Mexico. The highest mountains are in the Trans-Mexico Volcanic Belt, with numerous summits above 4,000 meters, and up to about 5,640 m. There are many different kinds of soils in Mexico (Krasilnikov et al. 2013), but no descriptions of ultramafic soils were found in published documents, other than for soils on the Vizcaíno Peninsula (Alexander 2007, Alexander 2019). Surface soils from sites with ultramafic strata in Guanajuato, Puebla, and Oaxaca (Fig. 11-1) were evaluated to learn how the plant distributions and the concentrations of chemical elements in the leaves were related to soil properties (Navarrete Gutiérrez et al. 2018).

A. Geology

South of the North American craton, the oldest terranes that are now in Mexico and Central America are the Oaxaquia, Maya, and Chortis terranes that were associated with Amazonia in the Proterozoic (Keppie 2004). Following the break up of Rodinia in the Proterozoic eon, the Mixteca and Sierra Madre terranes were added to South America, in association with the Oaxaquia, Maya, and Chortis terranes. As part of Gondwana, these terranes were rotated to a position opposite Laurentia (ancient North America). With the formation of Pangea, they were attached to Laurentia and have been a part of North America since the Mesozoic era.

Most of the ultramafic rocks in Mexico are in terranes west of the post-Pangea core, but with minor ultramafic bodies in Oaxaquia and Mixteca terranes, and in the Cuicateco terrane that was derived from Oaxaquia and the Maya terranes (Centeno-García 2017). Westward from the post-Pangea terranes are the parautochthonus Guerrero composite terrane (Centeno-García et al. 2011), the allochthonous, oceanic Vizcaíno terrane, and the Alisitos terrane. Origin of the Alisitos terrane is not completely resolved, but its development may be intermediate between the Guerrero composite and Vizcaíno terranes, with both parautochthonous and allochthonous components. Camprubi (2017) has shown how the Phanerozoic terranes of northwestern Mexico are related to terranes west of the North American craton farther north. Oaxaquia is more closely related to the Proterozoic Grenville province, which is exposed in central Texas (Chapter 6).

B. **Climate**

Mexico has wet to dry subtropical climates. The wettest climates, with annual precipitation from 2000 mm up to about 4000 mm, are on lowlands near the Gulf of Mexico and the Isthmus of Tehuantepec and on high mountains in the Sierra Madre del Sur. The drier climates are on the north, near the Horse Latitudes, or subtropical high pressure zone, and west of high mountains that intercept the prevailing winds from the east and deplete the moisture in them as they pass westward. Most of the precipitation is through summers, but during winters in Baja California north of the southern tip of the peninsula. Also, much of the rainfall at El Arco (Fig.11-2), near Calmalli (Fig. 11-1), is from late summer to autumn tropical cyclones (Hastings and Turner 1965).

C. Ultramafic Sites in Southern Mexico

Navarrete Gutiérrez et al. (2018) investigated the soils and plant leaf contents at three locations in Mexico to learn of nickel distributions in the soils and plants of some ultramafic areas. They sampled surface soils (0-30 cm depths) and plants from sites in Guanajuato, Puebla, and Oaxaca (Fig. 11-1). The ultramafic materials were highly serpentinized peridotite at the southern sites and less serpentinized pyroxenite at the Guanajuato sites.

The Oaxaca, Puebla, and Guanajuato sites were at about 1800, 1200 m, and 1900 m asl. Precipitation at the Oaxaca location is about 717 mm annually and the vegetation was a deciduous forest with oak and pine trees. Precipitation at the Puebla location is about 530 mm annually and the vegetation was a deciduous forest with cacti. Precipitation at the Guanajuato location is about 559 mm annually, or 698 at the weather station (Fig 11-2), and the sites investigated had low plant diversity in an area dominated by oak forest and grassland.

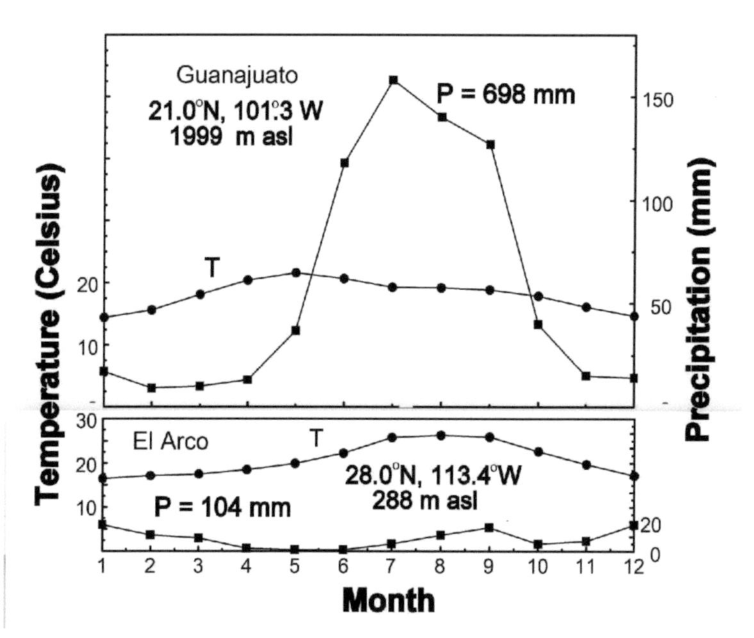

Figure 11-2. Mean monthly temperatures and precipitation at Guanajuato and El Arco. Calmalli is near El Arco..

Results from the soil analyses show high values of Cr, Co, and Ni, as expected in ultramafic soils, for sites at the Oaxaca and Puebla locations, but not for the sites at Guanajuato (Table 11.1). Nevertheless, caracus wigandia (*Wigandia urens*) from a Guanajuato site had the highest leaf concentrations of Co (7.4 µg/g) and Ni (43.1 µg/g) among 24 plants analyzed for these elements. Uncharacteristically for ultramafic soils, some of them had more exchangeable Ca than Mg, and the soils with the greatest proportions of Ca had the highest pH values. Navarrete Gutiérrez et al. (2018) did not mention any possible sources for the extra Ca, which may be from nonultramafic sources associated with the ultramafic soil parent materials.

Table 11.1. Surface soil analyses from the report of Navarrete Gutiérrez et al,.(2018).

Location (number of samples) [a]	Organic Carbon	Total N	CEC	Exchange Cations			pH	Total Element		
				Ca	Mg	K		Cr	Co	Ni
	g/kg		mmol/kg					mg/kg		
Oaxaca (2)	59	2.0	215	36	143	36	6.5	1755	132	2610
Puebla (2)	30	1.4	172	78	42	–	7.6	1129	118	2105
Guanajuata (2)	23	1.0	351	119	159	58	7.0	305	48	205
Guanajuata (1)	11	0.1	432	217	152	48	8.3	142	45	192

[a] Only data from sites with soil sample pH differences less than 1.0 were paired (n=2) in this table.

D. Alisitos Terrane

The Calmalli ophiolitic body near El Arco, Baja California Norte, is a representative of the Alisitos terrane. It contains imbricated slices of serpentine consisting of foliated antigorite that surround peridotite, pyroxenite, and amphibolite (Radelli 1989). The ophiolite and serpentine are strongly folded and metamorphosed to greenschist facies. Peridotite and serpentinite are so intricately folded and faulted along with metabasalt and other rocks in the "ophiolite of Calmalli" (Radelli 1989), that it was difficult to find a large block of ultramafic rock to describe a complete ultramafic landscape (Alexander et al. 2007a).

Table 11.2. Calmalli ultramafic soil properties from pedon descriptions, and pH or effervescence (ef) of fine-earth (soil <2 mm).

Horizon		Munsell Color		Texture (feel)	Struc- ture	Cons.	Roots	Bndr	pH
sym.	depth	moist	dry			dry			or ef
Pedon CO2, Lithic Haplargid, summit of hill, slope 5%, 40% Gr, 8% Cb, St<1%, 1% rock outcrop									
A	0-4 cm	5YR 4/4	6/5	vGrSL	1vfsbk	s	nil	as	7.0
Bt1	4-10	5YR 4/6	6/6	vGrSCL	2vfsbk	sh	2vf, f	cw	7.1
Bt2	10-27	5YR 4/6	5/6	vGrCL	2vfsbk	h	2f, m	ai	7.2
R	27-30+	highly fractured bedrock, somewhat schistose							
ped coatings; thin, discontinuous in Bt1, continuous in Bt2 horizon									
Pedon CO4, Lithic Haplargid, upper hill slope, 9% convex, 8% Cb, I% St, 9% rock outcrop									
C	3-0 cm	Gr>80%, Cb 8%, St 1%; yel, green, black serpent., some pale red rock fragments							
A	0-3	7.5YR 4/4	6/5	vGrSL	0 mass	s	nil	as	slt ef
Bt	3-14	7.5YR 4/4	6/5	vGrSCL	1fsbk	sh	1vf,f, v1m	ai	no ef
R	14-18+	highly fractured bedrock, serpentinite							
ped coatings, Bt horizon, thin, discontinuous in tubular pores, sparse on ped faces									
Pedon CO3, Typic Haplargid, lower hill slope, 34% concave, 27% Cb, 3% St, no rock outcrop									
C	2-0 cm	Gr>60% , Cb 27%, St 3%; cobbles and stones 5-2.5YR 5/4 on surfaces							
A1	0-3	7.5YR 4/3	6/4	vCbSL	1copl	s	v1vf	cs	no ef
A2	3-9	7.5YR 4/3	6/4	vGrSL	1msbk	sh	1vf, f	cw	slt ef
Btk1	9-30	7.5YR 4/3	6/4	vGrCL	2fsbk	h	1m,1f	gs	str ef
Btk2	30-72	7.5YR 4/3	6/3	vGrCL	2vfsbk	h	v1f	ai	m ef
R	72-75+	highly fractured bedrock, serpentinite							
ped faces 7.5YR 4/4 dry, ped coatings: Btk1, thin discontinuous in tubular pores and sparse moderately thick on ped faces; Btk2, thin discontinuous in tubular pores and common moderately thick on ped faces pebble and cobble coating, white: Btk1, discontinuous on lower sides of pebbles and cobbles Btk2, continuous white coatings on all sides of pebbles and cobbles									

Abbreviations: texture - v, very; Cb, cobbly; Gr, gravelly; C, clay; L, loam or loamy; S, sand or sandy; structure - 0, structureless; 1, weak; 2, moderate; co, coarse; f, fine; m, medium; mass, massive; gr, granular; pl, platy; sbk, subangular blocky; consistence, cons. - h, hard; s, soft; sh, slightly hard; vh, very hard; roots - 1, few; v1, very few; f, fine; m, medium, vf, very fine; boundary, Bndr - a, abrupt; c, clear; d, diffuse; g, gradual; I, irregular; w, wavy; effervescence, ef, - slt, slight; m, moderate; str, strong.

Aridisols (Haplargids) dominate the ultramafic landscapes. The more shallow Haplargids have Bt horizons and deeper Haplargids on colluvial footslopes have Btk horizons, but not enough lime for calcic horizons. Three pedons were chosen to represent a sequence of ultramafic soils from the summit of a hill down the side of the hill to lower slope positions (Table 11.2). The soils were shallow on the summit, very shallow near midslope, and moderately deep on the lower slope, as might be expected on a surface formed by the runoff of rain water.

Serpentine was a dominant mineral in the soils, but subangular to rounded quartz grains were common in the fine and very fine sand fractions, and there were some feldspars and sparse mica along with the quartz. The quartz was not evident in the coarser sand fractions. It may be from nonultramafic inclusions in the parent material, or it may be aeolian. Only small fractions of the fine sand grains were heavy (SG>2.89), including magnetic grains (presumably magnetite). Other heavy minerals were mainly Clinopyroxenes, orthopyroxene, amphiboles, and some olivine. The abundance of hornblende and sparsity of tremolite and actinolite, which are generally the more common amphiboles in ultramafic soils, may be indicative of a nonultramafic source of amphiboles in the soil parent materials. Magnesite veins in the rock strata and cementation of alluvial fill in arroyos with silica are conspicuous phenomena in the ultramafic terrain.

Torote colorado (*Bursera microphylla*) was present on the summit and midslope soils and torote blanco (*Pachychormus discolor* var. *pubescens*) was common on the lower slope. Palo verde (Parkinsonia microphylla) was sparse on the summit and present on the midslope. Burrobush (*Ambrosia dumosa*) was a common shrub on the summit soil and brittlebush (*Encelia farinosa*) was common on the midslope and lower slope. Ocotillo (*Fouquieria splendens*) was common on the upper hill slope. A spurge (*Euphorbia xanti*) was present on the summit soil. Agave (*Agave* sp.) was common on the summit soil. Pitayita (*Echinocereus engelmannii*) was common on the summit soil and pitaya agria (*Stenocereus gummosus*) was common on the upper hill slope. Also, other cacti were present, but sparse, in the *Cylindropuntia*, *Ferocactus*, and other genera. Cardon (*Pachycereus pringlei*), a prominent cactus in Baja California, was present on nonultramafic soils of the Calmalli ophiolitic body.

12 Caribbean Plate

The Caribbean is a tropical area corresponding closely to the Caribbean plate and its immediate surroundings. From a geological perspective the area would include the Greater Antilles on the north, the Lesser Antilles on the east, the northern margin of South America, and practically all of Central America. Ultramafic terranes of the Greater Antilles, Cuba, and Central America are considered in this chapter, but none of the ultramafic terranes in South America..

A. Geology

The geological origin of the Caribbean plate has been controversial. An origin summarized by Mann (2007) is a commonly accepted scenario. According to this hypothesis, a volcanic arc that formed in the Pacific Ocean passed over a mantle plume, possibly related to the Galapagos hot spot, and a large oceanic platform formed there during the Mesozoic era. Subsequently, the arc and platform drifted eastward through a gap that developed between North America and South

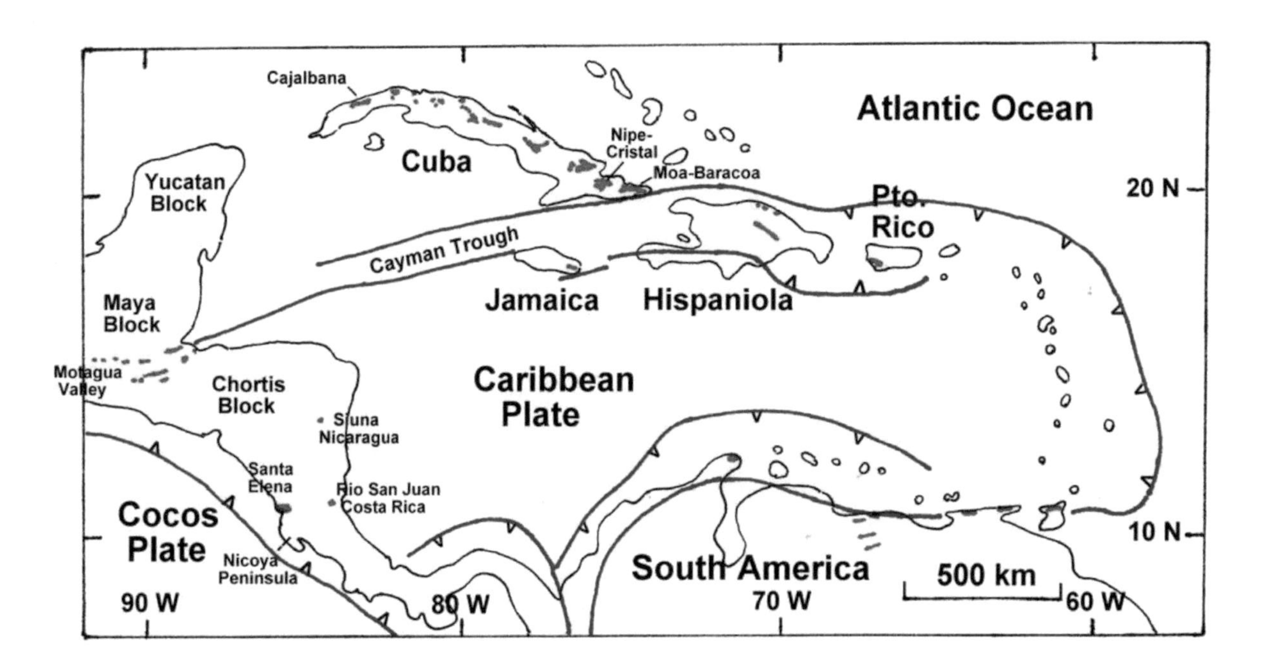

Figure 12-1. Locations of ultramafic rocks (red) in the Caribbean area.

America after Laurasia and Gondwana had separated in the breakup of Pangea. Remnants of an old volcanic arc are the roots of Cuba, and the Mesozoic ocean platform is now a large part of the Caribbean seafloor.

Exposures of ultramafic rocks are concentrated around the margins of the Caribbean Plate from Guatemala to Puerto Rico and along the northern margin of Southern America (Fig. 12-1). Also, there are several exposures of ultramafic rocks in Nicaragua and Costa Rica. Picrites and komatiite are found near the trailing edge, or Pacific margin, of the Caribbean Plate. They are on Gorgona Island, Colombia (Kerr et al. 1996), and possibly in the Nicoya complex, Costa Rica (Alvarado et al. 1997). These volcanic rocks have been dated at 88-90 Ma and they attest to high temperatures that may have been related to the Galapagos hotspot.

B. Climate.

The climate at the latitude of the Caribbean Plate (about 10 to 22°N) is hot and ranges from very moist to seasonally dry, and locally very dry. Prevailing winds in this area between the subtropical high, or "calms of Cancer", and the Intertropical Convergence Zone are from the east, or northeast. There are two rainy seasons, one in May to June and another in September to November that occur when the sun passes overhead, or soon thereafter. The main dry season is from January to April, when the sun is over the southern hemisphere. Following the dry season, prevailing winds carry moderate amounts of moisture to eastern or northeastern shores and large amounts to higher elevations, but they are drier on the southwest of the uplands. The contrast is evident between Bluefields on the Caribbean coast of Nicaragua with a mean annual precipitation of 4200 mm and Liberia about 235 km southwestward across to the Pacific side of Costa Rica with 1661 mm (Fig. 12-2).

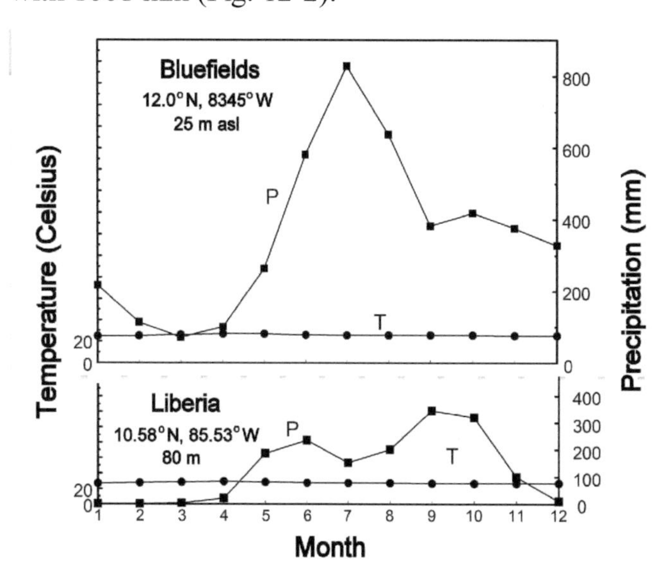

Figure 12-2. Mean monthly temperatures and precipitation on the Caribbean (Bluefields) and Pacific (Liberia) sides of Central America.

Ravalo et al. (1986) have produced monthly isohyetal maps for Puerto Rico that show how the annual patterns of the precipitation distribution can by very complex (section E3 in this chapter). Cyclones are large sources of rain water on the Caribbean islands during late summer to autumn. They commonly move westward across the Atlantic ocean and turn toward the northwest as they approach Puerto Rico and the other Caribbean islands. Weather stations on the arid south coast of Puerto Rico have occasionally recorded autumn rainfall >200 mm/day (Southeast Regional Climate Center).

C. Nicaragua and Costa Rica

There are small areas of ultramafic terrain in Nicaragua and Costa Rica and a large area of about 340 km^2 on the Santa Elena Peninsula of northwestern Costa Rica (Fig, 12-1). Only the Santa Elena Peninsula has enough ultramafic soils and vegetation information to warrant mention here. A broad (small scale) ultramafic geoecology survey and report of the area was made for the Area de Conservación Guanacaste (Alexander 2008, 2018, unpublished, accessible from *Soils and Geoecology* online).

There is a long December to April dry season on the Santa Elena Peninsula, as at nearby Liberia (Fig. 12-2), also. The soils are seasonally dry (ustic SMR). Most of the area has been severely dissected. The ultramafic soils are mainly very shallow Entisols, shallow to very deep Inceptisols, and moderately deep to very deep Alfisols. The very deep soils are in colluvium. The vegetation is mostly savanna (Fig. 12-3) and some scrubby woodland. The most common plants are nancite (*Byrsonima crassifolia*), roupale (*Roupala crinervis*), turnera (*Turnera pusilla*), and crinkleawn (*Trachypogon plumosa*). Jaragua (*Hyparrhenia rufa*), a common invasive grass, commonly avoids ultramafic soils.

Figure 12-3. Ultramafic landscapes of the Santa Elena Peninsula. A. A woodland-savanna with crinkleawn, nancite and roupale trees, and turnera shrubs on a shallow Inceptisol. B. Maria Marta Chavarria recording plant species in a crinkleawn savanna on Alfisols.

Above about 500 m on the roadless and practically inaccessible western part of the Santa Elena Peninsula there is a relatively moist, often cloud engulfed area of several hundred hectares with dense dwarf forest. Dauphin and Grayum (2005) found abundant corticolous (bark) bryophytes on trees of the cloud forest in contrast to the more usual soil and rock bryophytes of the seasonally dry, semideciduous forests, or woodlands.

There is one serpentine endemic plant (*Simsia santarosensis*). Five plant species occur in Costa Rica only on ultramafic soils: *Simsia santarosensis* (Asteraceae), *Agave seemanniana* (Agavaceae), *Bursera schlechtendalii* (Burseraceae), *Melocactus ruesti* (Cactaceae), and *Schizachyrium malacostachyum* (Poaceae). None of the plants on ultramafic soils are Ni- hyperaccumulators; the highest Ni concentration was found to be 275 mg/kg in *Buchnera pusilla* (Reeves et al. 2007).

D. Cuba

The lithological units of Cuba are a complex amalgamation of volcanic arc and sea floor strata from the Caribbean Plate mixed with Mesozoic to Paleogene sedimentary strata of the southern margin of the North American Plate and with a cover of Neogene sediments (Pardo 2009). There is about 7500 km^2 of ultramafic rock terrain in Cuba (Borhidi 1991).

Cuba has large areas of ultramafic rocks on both old and young land surfaces, and diverse climates. The old land surfaces, which are on the Cajálbana tablelands on the west and in the Nipe-Cristal and Moa-Baracoa areas at the east end of the island (Fig. 12-1), have been exposed for many millions of years; and the younger surfaces, presumed to be of Quaternary age (Borhidi 1991), are nearer the center of Cuba. The diverse geology and climates have led to large numbers of plant species, and because of isolation on the island, there are large numbers of endemic plant species. Borhidi (1991) identified 750 serpentine endemics found only on the 4800 km^2 of older surfaces, 128 endemics found only on the 2700 km^2 of younger ultramafic surfaces, and 42 endemic species common to both areas.

The ultramafic soils of Cuba are hot (isothermic STR) or very hot (isohyperthermic STR) and moist (udic SMR) or seasonally dry (ustic SMR); and there are minor areas of arid soils. A general soils map of Cuba (Marrero Rodríguez et al. 1989) shows ultramafic Ferralicos (Púrpura) and Pardos sin Carbonatos in the eastern (Oriente) region, mainly south of the Bahia de Nipe (Nipe-Cristal area) and south and southeast from Moa (Moa-Baracoa area). Ferralicos are prevalent ultramafic soils in the Cajálbana area of tablelands (Fig. 12-1), also. Rojo Pardusco Ferromagnesial and Pardos sin Carbonatos are the most common ultramafic soils in the Central regions from Habana to Holguin. The Ferralicos (Púrpura) appear to match most closely to Ferralsols (World Reference Base), or Oxisols (US Soil Taxonomy); the Pardos sin Carbonatos correspond to Phaeozems, or Mollisols; and the Rojo Pardusco Ferromagnesial soils pair to Cambisols, or Inceptisols.

Ségalen et al. (1980, 1983) sampled old ultramafic soils that are typical of those on the Cajálbanal tablelands. The soils have high clay contents. The order of clay mineral abundance was reported as goethite > hematite > kaolinite = gibbsite in the surface and goethite >> hematite = kaolinite > gibbsite in the subsoil. The soils are Ferralsols, or moist Oxisols. Similar soils have been found in the Oriente region, with halloysite rather than kaolinite where the soils are wetter, but Cerpa et al. (1999) did not find either kaolinite or halloysite in Oxisols at three sites there, and they reported maghemite, rather than hematite, along with abundant goethite.

Dirven et al (1976) sampled three soils in a toposequence that appear to be representative of soils in the Central region of Cuba. From higher erosional surfaces to a lower surface they sampled Chromic Phaeozems (or lithic, haplic Mollisols), Chromic Cambisols (or Inceptisols), and Pellic Vertisols (or seasonally wet Vertisols). Clays in the well-drained soils were reported to have high goethite and low hematite contents, with moderate kaolinite(?) and low smectite in surfaces and moderate smectite and low kaolinite(?) in subsoils. Clays in the poorly drained soils in lower landscape positions were reported to be predominantly smectite. Silica released by the weathering of serpentinite and not used to produce smectites accumulates as quartz in the well drained soils. Some of the Mg leached from the ultramafic soils in higher landscape positions produces magnesite ($MgCO_3$) in the poorly drained Vertisols; and reduced Fe produces siderite in the Vertisols (Dirven et al. 1976).

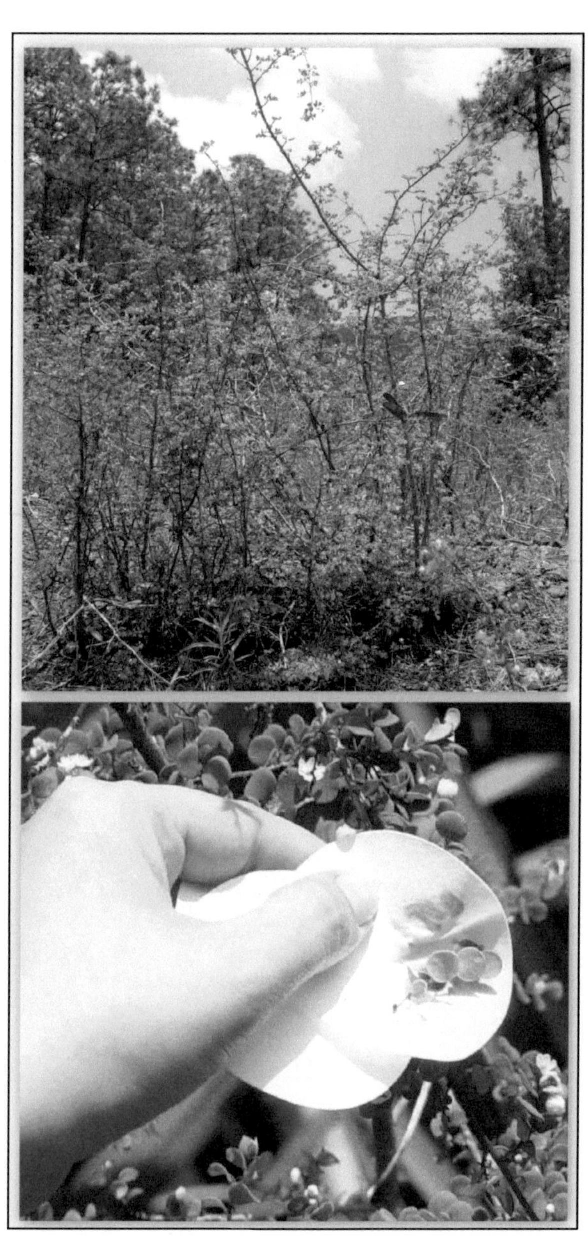

Figure 12-4. *Phyllanthus orbicularis*, a Ni-hyperaccumulating shrub common in Cuba. Nickel from leaves pressed onto a filter paper saturated with dimethylglyoxime forms a red precipitate. Photos by Micheal Davis.

Shallow, or very shallow, ultramafic soils are common in some parts of the Central region, as reported for the Sabana de Comagüey by Bennett and Allison (1928). These ultramafic soils have relatively high clay contents.

Ecologists commonly remark that the plants on ultramafic soils are generally more xerophytic than the those on nonultramafic soils. Borhidi (1991) compared the leaf areas of large numbers of species over broad temperature and moisture gradients. On ultramafics, granodiorite, and limestone, leaf areas (based on a leaf area index) increased linearly with temperature and precipitation. The index indicated smaller leaves for plants on ultramafic soils than for those on granodiorite or limestone soils with comparable temperature and precipitation.

Thorniness, as judged by the proportion of thorny taxa, was low on limestone and very low on serpentine soils with precipitation >2000 mm/year. It increased exponentially to >30% as the precipitation declined to <1000 mm/year.

Although Borhidi (1991) has demonstrated the xerophytic aspects of vegetation on ultramafic soils, his reasons for the xerophytic nature of serpentine plants are less convincing, or based on common misconceptions about ultramafic soils. More convincingly, Berazaín (2001) emphasized nutrient deficiencies, or imbalances, and/or toxicities as the primary causes of the distinctive aspects of serpentine vegetation.

Motito Marin et al. (2004) have noted the diverse taxa of bryophyte flora in the ultramafic soil areas of Cuba. There were 228 taxa of mosses and 354 taxa of liverworts. Although 38% of the species were endemic to Cuba, none were serpentine endemics.

Cuba is noted for the large number of Ni hyperaccumulating plant species there (Reeves et al. 1998). One hundred and seventy-three (173) Ni-hyperaccumulators (including *P. orbicularis*, Fig. 12-4) were recognized by the beginning of the 21st century (Berazaín 2004). The families with the most Ni-hyperaccumulators are Euphorbiaceae, Asteraceae, Buxaceae, Rubiaceae, and Myraceae (Berazaín 2004). The monocotyledons are poorly represented by Ni-hyperaccumulator species.

The Nipe soil, which was first identified in Cuba (Bennet and Allison 1928), has been the typical example of an ultramafic Oxisol in the Caribbean area. Its features are described in the Appendix (IV).

E. Northern Margin of the Caribbean Plate

Appreciable amounts of ultramafic rocks are exposed in Guatemala, Puerto Rico, and Hispaniola, and there are some in Jamaica. They are mostly serpentinized harzburgite and lesser amounts of dunite and serpentinized lherzolite (Lewis et al. 2006).

12E1 Guatemala

In Guatemala the ultramafic rocks are associated with the Motagua Valley, which runs from west to east along the northern edge of the Caribbean Plate (Fig. 12-1). The ultramafic rocks were emplaced by thrusting to the north over the Polochic fault onto the Maya block and to the south over Motagua fault onto the Chortis block (Lewis et al. 2006).

Simmons et al. (1959) reported the mapping of 222 km^2 of ultramafic soils in Guatemala. The soils are mostly shallow to moderately deep, seasonally dry Inceptisols (45.6 km^2); moist Inceptisols (32.9 km^2); seasonally dry to moist Alfisols (133.6 km^2); and humic Ultisols (9.7 km^2). An ultramafic Alfisol has been described in the area recently (Table 12.1, Fig. 12-5).

Table 12.1 An ultramafic soil representative of those in the Motagua Valley.

A pedon of the Guapinol series (Udic Haplustalfs) in complex with Lithic Haplustalfs (Fig. 12-5)
located off highway RD-4, about 1 km north of highway CA-9
15°23' N latitude, 89°02' W longitude, 250 m altitude
Departamento: Izabal, Mariscos 1:50,000 quadrangle

Landform: hillside, convex-concave slope 53% southeast (150° magnetic compass) slope
Parent material: slightly serpentinized peridotite
Surface rock and stoniness: rock outcrop <1%; 6% "stones", 18% cobbles
Vegetative cover: pine/shrub savanna, with a closed angiosperm forest on an adjacent N-facing slope
Oi nil
A 0-5 cm dark reddish brown (5YR 3/4, 4/5 dry) very gravely loam; strong, medium granular structure; hard, friable, and slightly sticky and slightly plastic; 6% "stones", 18% cobbles, 30% gravel; common very fine and few fine roots; neutral (pH 6.8); abrupt, smooth boundary
Bt1 5-30 reddish brown (5YR 3/5, 4/6 dry) very cobbly clay loam; strong, medium, subangular blocky; very hard, firm, and sticky and plastic; continuous, distinct, dark reddish brown (5YR 3/3) clay coatings on ped faces; 9% "stones",18% cobbles, 15% gravel; common fine, few medium, and very few coarse roots; neutral (pH 6.9); diffuse boundary
Bt2 30-56 reddish brown (2.5-5YR 4/6, 4/6 dry) very stony clay loam; moderate, medium, subangular blocky; very hard, firm, and very sticky and plastic; many, faint, dark reddish brown (5YR 3/4) clay coatings on ped faces; 18% "stones", 12% cobbles, 10% gravel; few fine and medium and very few coarse roots; neutral (pH 7.0); abrupt, irregular boundary
R 56-100+ hard, moderately fractured peridotite bedrock, serpentine along fractures; thick reddish yellow (5YR 6/8) weathering rims with diffuse gradation to dark greenish gray (5G 4/1) unweathered rock
Notes: described January 1, 2008 by E.B. Alexander; pH by bromthymol blue (BTB) indicator; bedrock weathered to at least 3 or 4 m depth
Soil class: loamy-skeletal or clayey-skeletal, magnesic, isohyperthermic Udic Haplustalf
World Reference Base (WRB) classification: Luvisol

Hor	Depth cm	Text. (feel)	Ca	Mg	Ca/Mg molar ratio	EA	CEC	BS %	pH
			mmol+/kg			mmol+/kg			BTB
A	0-5	vGrL	33	215	0.15	116	364	68	6.8
Bt1	5-30	vCbCL	16	285	0.06	115	416	72	6.9
Bt2	30-56	vStCL	9	480	0.02	118	607	81	7.0

Note: extraction of cations with 0.5 molar KCl and titration with EDTA.
EA, Exchange acidity (pH 8.2) by KCl-triethanolamine method. BS, pH 8.2 base saturation.

Figure 12-5. The soil and landscape of a hot, seasonally dry Alfisol in a Caribbean pine (Pinus caribaea) savanna of the Motagua Valley, Guatemala.

Nickeliferous laterites are present in six complexes on the north side of the Motagua Valley (Lewis et al. 2006, Valls 2006). A notable feature of the Ni-laterites in Guatemala is the high magnetite concentration, up to 30%, in the saprolite (Valls 2006).

12E2 Hispaniola

Ultramafics **of** the Loma Caribe area in the Dominican Republic occur on a narrow, 95 km long body of serpentinized peridotites (Fig. 12-1) that were exposed to weathering and erosion in the early Miocene (Proenza et al. 2008). There has been plenty of time for the development of Ni-laterites. Garnierites in the saprolite have been described and characterized in several recent articles. The garnierites occur in fractures, or veins, and as boxwork (Proenza et al 2008). Minerals in the garnierite are mainly nickeliferous lizardite and an intermediary between hydrous talc ("kerolite") and hydrous willemseite; there are no smectites (Proenza et al. 2009, Villanova-de-Benavent et al. 2011). The garnierite contains 1.2 to 2.3% (12000-23000 ppm) Ni. A nickeliferous sepiolite found in the garnierite at the Falcondo mine of Loma Caribe has been called falcondoite.

Berazaín (2004) listed two Ni-hyperaccumulators for Hispaniola; a *Phyllanthus* sp. in the Phyllanthaceae and a *Psychotria* sp. in the Rubiaceae.

12E3 Puerto Rico

Puerto Rico is oriented east to west, with a central range of mountains, or Cordillera Central, from near Mayagüez to the center of the island, a Sierra de Capey on the southeast, and a

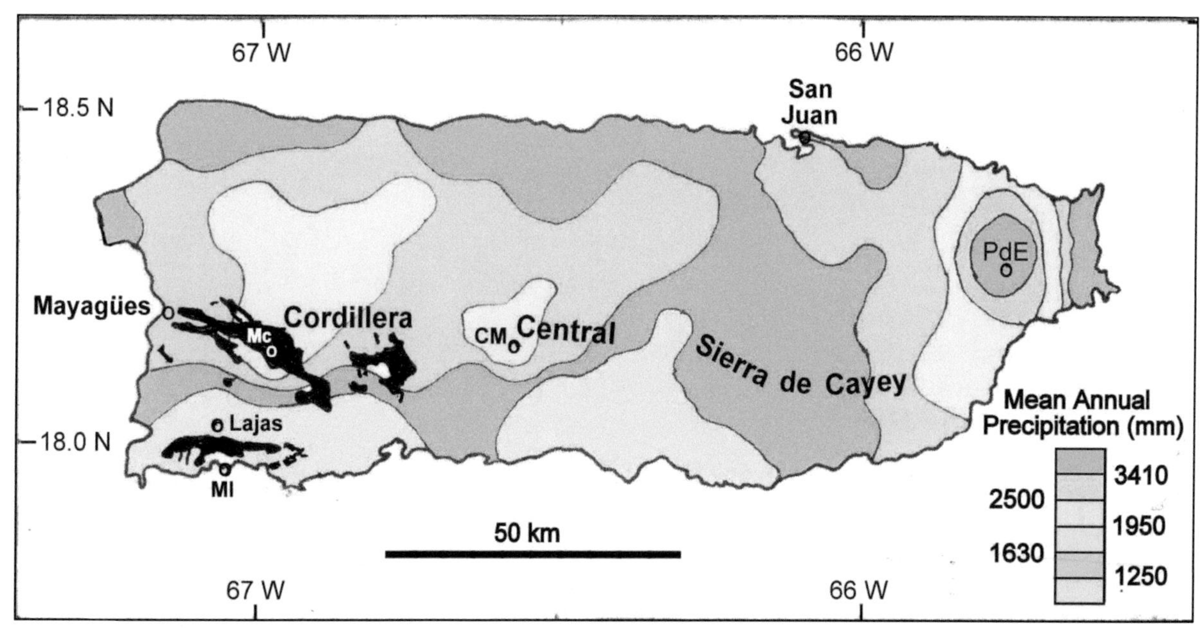

Figure 12-6. Puerto Rico. An isohyetal map, with the locations of ultramafic terranes in southwestern Puerto Rico after Gomez-Gomez et al. (2014). CM, Cerro Maravilla; Lj, Lajas; Mc, Maricao; MI, Magueyes Island; PdE, Pico del Este.

Sierra de Luquillo more than 1000 m high on the northeast (Fig. 12-6). Cerro de Punta (1327 m) in the Cordillera Central is the highest mountain.

Isohyetal maps for each month show the patterns of precipitation distribution across Puerto Rico (Ravalo et al. 1986). Any months with rainfall <75 mm may be considered months of drought at the tropical temperatures of Puerto Rico (Walter 1973). The Sierra de Luquillo has rainfall >100 mm in February and March, which are the driest months throughout Puerto Rico. Otherwise, only the higher areas in the Cordillera Central and northern slopes toward the coast have mean rainfall >75 mm in these months. The area from Lajas and Sabana Grande southward has mean rainfall <50 mm from December to March, and some coastal areas have mean rainfall <25 mm/month. These are areas that have soils with ustic (periodically dry) and aridic moisture regimes. The west coast is unique in that prevailing winds from the west cause the area around Mayagüez, especially north of Mayagüez, to be very rainy from May to September or October.

Table 12.2 Precipitation, mean annual, for stations located in Figure 12-6.

Weather station Puerto Rico	Long. degr. W	Latitude degr. N	Altitude (m asl)	Rainfall (mm)	Wettest months
San Juan	66.06	18.24	14	1510	Oct, Nov
Punto de Este	65.75	18.27	1052	4345	May, Nov
Cerro Maravilla	66.55	18.15	1204	2535	Sep, Oct
Maricao 2 SSW	66.99	18.15	863	2421	Sep, Oct
Mayagüez	67.13	18.20	8	1733	Aug, Sep
San Germán	67.05	18.08	12	1459	Sep, Oct
Lajas	67.07	18.03	(10)	1218	Sep, Oct
Isla de Magueyes	67.05	17.97	8	743	Jun, Sep

Precipitation data for at least 25 years, 65 years for most stations, 1948-2012.

Mesozoic serpentinized peridotite of the Caribbean Plate has been thrust over and imbricated with younger rocks from the Cretaceous through the Paleogene (Laó-Davila et al. 2012) and exposed near the western end of the Cordillera Central and in hills south of the Cordillera. The ultramafic rocks are in the Monte de Estado area at the western end of the Cordillera Central (Fig. 12-1), on the Sabana Alta of Cabo Rojo, on the Cerros de las Mesas de Mayagüez, on the Lomas de Santa Marta in the Guanajibo Valley, and in the Sierra Bermeja south of the Lajas Valley (Fig. 12-6).

About 13 000 ha of ultramafic soils have been mapped in Puerto Rico. Approximately 75% of them were mapped as Mollisols and 25% were mapped as Oxisols, and a very small area of Alfisols was mapped on the Lomas de Santa Marta, near San Germán. Ultramafic soils of the Monte de Estado, Cerros de las Mesas, and Sabana Alta areas are predominantly moist, shallow, argillic Mollisols on steeper slopes and moist Oxisols of the deep Nipe and moderately deep (shallow for Oxisols) Rosario series on lesser slopes (Fig. 12-7). Ultramafic Oxisols near the crest of the Cordillera Central are very moist (Cerro Gordo series). Some of the ultramafic Mollisols at low elevations may be periodically dry, but they were not perceived to be dry enough to have ustic soil moist regimes.

Figure 12-7. Ultramafic soils and vegetation in Puerto Rico. A. A seasonally dry landscape with Mollisols and with some of the trees and shrubs listed in Table12.4: doncella, palo de cucubano, capa prieto, caoba, manzanilla del monte, carrasco, tintillo, and grass (lamilla, *Boutelua* sp.). B. Deep and moderately deep ("shallow") Oxisols. C. A very moist Oxisol, Cerro Gordo series.

Although the soils are moist even at low elevations along the west coast at Mayaguéz and Cabo Rojo, soils of the Sierra Bermeja, which is along the south coast, less than 20 km south from Cabo Rojo, were considered to have aridic soil moisture regimes (Lugo-Camacho 2008). There are small outcrops of ultramafic rocks in the Sierra Bermeja, but not enough ultramafics to map any ultramafic soils there.

Laboratory analyses for some of the ultramafic soils have been published in a report of Mount and Lynn (2004): Nipe series (deep Oxisols), La Taina series (labeled Maresua in the report, shallow argillic Mollisols), Rosario series (moderately deep Oxisols), and Santa Marta series (moderately deep Alfisols). The Oxisols have the features of Ni-laterites, as described in Appendix IV. Red iron oxides from the Nipe soils form enduring stains (Fig. 12-8).

The physical properties of the Nipe soils are quite favorable for some plants, but the P-absorption capacity is very high compared to other Oxisols in Puerto Rico (Fox 1982). Although the Nipe soils support luxuriant forests, they are not suitable for cultivated crops. Undisturbed Nipe soils are friable and they have high available water capacities. Lugo-Lópes et al. (1981) measured the rates of infiltration into the soils of eight soil orders in Puerto Rico after they were saturated for eight hours. Infiltration was most rapid in Oxisols, and the rate for the Nipe soil, at 15.6 cm/hr, was one of the highest rates. Lohnes and Demirel (1973) determined some physical properties of ten old soils, including Nipe (Table 12.3). Its liquid limit is low for a clayey soil, and the plastic limit is high, resulting in low plasticity indices in both surface and subsoil. Even with moderate bulk density, the porosity was high, because the relatively high bulk density was related to high Fe content and high particle density. The matrix of a Nipe soil described by Eswaren et al. (1979) was plasmic, lacking distinctive features other than ferruginous nodules and remnants of relict ferriargillans.

Figure 12-8. The author and Manuel Matos (NRCS, Puerto Rico) with hands red from handling Nipe soil. Photo by Dr. Rebecca Corbala.

Table 12.3 Some physical properties of a Nipe soil reported by Lohnes and Demirel (1973).

Hor.	Depth (cm)	Clay (%)	Liquid Limit	Plastic Limit	Bulk Density (Mg/m^3)	Particle Density (Mg/m^3)	Porosity (%)
Ap	0-25	44	48	39	1.27	3.17	60
B	74-95	72	43	36	1.41	3.27	57

Only three species were found in a brief study of earthworms at 0-10 cm depths in Nipe soils (Hubers et al. 2003).

Natural vegetation on the Oxisols is generally dense forests of large trees. On the seasonally dry Mollisols, there are forests that commonly contain shrubs, also, with slender grama (or lamilla, *Bouteloua repens*) and other grasses below or among the trees and shrubs. Common trees and shrubs mentioned in the ultramafic soil series descriptions are listed in Table 12.4. A tree fern (*Cyathea* or *Alsophila* sp.), was present only on very moist (perudic SMR) ultramafic Oxisols (Cerro Gordo series), along with palma de sierra (*Prestoea montana* var. *aculinata*). Much of the forest on the moist Oxisols along the coast near Cabo Rojo has been replaced by guinea (*Urochloa maxima*) and other introduced grasses in the past few hundred years.

Forty phanerogram (seed producing) species are restricted to ultramafic soils in Puerto Rico, although some of them may grow in nonultramafic soils elsewhere, and twelve of those species are endemic to Puerto Rico (Cedeño-Maldonado and Breckon 1996). None of the 12 species that are endemic to Puerto Rico are common plants; they are in ten different families, with three of the species in the Myrtaceae.

Medina et al. (1994) found two plant species that accumulated nickel in their leaves;70 µg/g in marbletree (*Cassine xylocarpa*, Celastraceae) and 700 µg/g in white rosewood (*Chionanthus domingensis,* Oleaceae). One species each in the Flacoutiaceae, Violacae, and Oleaceae, have been reported to be Ni-hyperaccumulators (Brooks et al.1977, Berazaín 2004). Campbell et al. (2013) found a Ni-hyperaccumulator (cachimbo grande, *Psychotria grandis*, Rubiaceae) in Puerto Rico that in specimens on ultramafic soils had 1800 to 8200 µg/g of Ni in the leaves and on limestone soils 95 to 125 µg/g of Ni in the leaves.

Table 12.4 Common trees and shrubs on the ultramafic soils of Puerto Rico.

Plant Family	Species	Common Name
Anacardiaceae	*Comocladia dodonaea*	carrasco (s/t) M
Annonaceae	*Annona glabra*	monkey-apple (t) Ox
Arecaceae	*Prestoea montana* var. *aculinata*	palma de sierra
Asteraceae	*Chromolaena oderata*	cariaquillo **(s)** M
	Wedelia reticulata	manzanilla del monte **(s)**
Bignoniaceae	*Tabebuia haemantha*	roble cimarrón (t)
Boraginaceae	*Cordia alliodora*	capá prieto (t)
Clusiaceae	*Caryophyllum antillanum*	palo de Maria (t) Ox
	Clusia clusiodes	cupeillo (t) Ox
	Clusia grundlacii	cupey del altura (v/s) Ox
	Clusia minor	cupey del monte (v/s) M
	Mammea americana	mamey, mammee apple Ox
Cyrillaceae	*Cyrilla racemiflora*	cyrilla, palo colorado (t) Ox
Fabiaceae	*Hymenaea courbaril*	algaroba (t) M
	Leucaena leucocephala	(s/t)
	Pithecellobium unguis-cati	escambrón colorado (s/t)
Flacourtiaceae	*Casearia decandra*	wild honey tree (t) Ox
Gesneriaceae	*Gesneria pedunculosa*	arbol de Navidad (t) M
Lamiaceae	*Clerodendrum aculeatum*	escambrón blanco (s) Ox
Malphigiaceae	*Byrsonima spicata*	palo de doncella (t) M
Meliaceae	*Swietenia mahagoni*	mahagony, caoba (t)
Rubiaceae	*Guettarda scabra*	palo de cucubano (t)
	Randia aculeata	tintillo (s/t)
Verbenaceae	*Lantana* spp.	cariaquillo (s)

M, Mollisols; Ox, Oxisols; s, shrub; t, tree; v, vine.

Appendix I

Geologic Time Scale

Time-Stratigraphic Unit	Time* (Ma)	Time-Stratigraphic Unit	Time* (Ma)
Phanerozoic Eon		Paleozoic Era	
Cenozoic Era		Permian Period	299
Quaternary		Carboniferous Period	359
Holocene	0.012	Devonian Period	416
Pleistocene	2.6	Silurian Period	439
Neogene Period		Ordovician Period	488
Pliocene Epoch	5.3	Cambrian Period	542
Miocene Epoch	23.0	Proterozoic Eon	
Paleogene Period		Neoproterozoic Era	1000
Oligocene Epoch	33.9	Mesoproterozoic Era	1600
Eocene Epoch	55.8	Paleoproterozoic Era	2500
Paleocene Epoch	66.5	Archean Eon	
Mesozoic Era		Neoarchean Era	2800
Cretaceous Period	145.5	Mesoarchean Era	3200
Jurassic Period	199.5	Paleoarchean Era	3600
Triassic Period	251.0	Eoarchean Era	4000
Hadean (informal unit) 4.0-4.56 Ga			

* Age at the beginning of the unit.

Appendix II

Periodic Table of Chemical Elements

H	----------																He
Li	Be	----------										B	C	N	O	F	Ne
Na	Mg	----------										Al	Si	P	S	Cl	Ar
K	Ca	Sc	Ti	V	Cr	Mn	Fe	Co	Ni	Cu	Zn	Ga	Ge	As	Se	Br	Kr
Rb	Sr	Y	Zr	Nb	Mo	Tc	Ru	Rh	Pd	Ag	Cd	In	Sn	Sb	Te	I	Xe
Cs	Ba	Lanthanides															----------
----------		Lu	Hf	Ta	W	Re	Os	Ir	Pt	Au	Hg	Tl	Pb	Bi	Po	At	Rn
Fr	Ra	Ac	Th	Pa	U	Np	Pu	other Actinides									----------

The elements between the two columns on the left and the six columns on the right are called transition elements. There are three rows of them in the table, with 10 transition elements in each of the rows beginning with (1) scandium, Sc, (2) yttrium, Y, and (3) lutetium, Lu. There are 14 lanthanides, beginning with lanthanum (La), and 14 actinides, beginning with actinium (Ac). The actinides beyond plutonium (Pu, $z=94$) are extremely rare up to $z=98$ and those with $z > 98$ are not found naturally on Earth.

Appendix III

Rocks and Clay Minerals in Ultramafic Soils

Rocks may be good for lichens, but plants need more than rocks for normal growth. Rocks weather to yield sand (0.06-2 mm), silt, and clay (particles <0.002 mm, or <2 μm) and elements that form clay minerals. Clay and organic matter are the active components of soils. They account for the storage and exchange of most of the water and available plant nutrients in soils. Clay and clay minerals warrant special consideration; they are discussed in some detail.

A. Rocks and Clay Mineralogy

The most common minerals in rocks and soils are quartz and feldspars. They are silicates with framework structures consisting of silicon tetrahedra in which all oxygen ions at the corners of the tetrahedra are shared by other tetrahedra in the framework. Silicon tetrahedra in the framework structures are tightly bound, limiting the reactivities of the minerals. Ultramafic rocks lack quartz, and feldspars are seldom found in them.

The common primary ultramafic rocks are peridotite and dunite. Olivine and pyroxenes are the main minerals in them and chromite is a common accessory mineral (Chapter 2). Olivine and pyroxenes are readily weathered and do not persist in strongly weathered soils, and the amounts of chromite are generally minor in them. The common and ubiquitous metamorphic products from dunite and peridotite are serpentine, brucite, and magnetite (Chapter 2). Brucite is readily weathered and is not an important mineral in any soils. Serpentine is much less readily weathered than olivine and pyroxenes, and it persists as a major clay mineral in ultramafic soils. It persists mainly in coarse clay (0.2-2 μm) and not much in fine clay (particles <0.2 μm). Magnetite is resistant to weathering and persists in many different soil size fractions.

Other than serpentine, no common minerals are present exclusively in ultramafic soils. Chlorite, and less commonly talc, are produced by the alteration of pyroxenes (Chapter 2). Talc is a major mineral in soils derived from soapstone or from talc schist. Steatite, or soapstone, can be produced from serpentine (Deer et al. 1966), or from dolomite with added silica. Also, unique varieties of some clay minerals are found in old, strongly weathered ultramafic soils.

Oxide and carbonate minerals are common in many rocks and soils, and sulfides and sulfates are present in some of them. The common oxide minerals in ultramafic rocks and ultramafic soils are chromite, magnetite, goethite, hematite, magnesite, and maghemite. The Mg and Fe-carbonate minerals, magnesite, and less commonly siderite, occur in some ultramafic soils.

IIIA1 Clay minerals in ultramafic soils

Most of the minerals produced by the metamorphosis of olivine and pyroxenes, and most of the clay minerals in ultramafic soils, are phyllosilicates (layer silicates), or Fe-oxides or oxyhydroxides. The building blocks of phyllosilicate minerals are silicon tetrahedra and various octahedra (Fig. 2-3, III-1). Tetrahedra have four faces and four apices, or corners, and octahedra have eight faces and six corners (Fig. 2-3), with oxygen ions at all of the corners. The layers of phyllosilicates consist of superimposed sheets of silicon tetrahedra and of octahedra, with cations other than Si in the octahedral sheets.

Tetrahedra sheet. Tetrahedra that share oxygen ions at three corners in a plane form a tetrahedral sheet (Fig. III-1A). The cations in tetrahedra are in 4-fold coordination, which means that each cation is linked directly to four adjacent oxygen ions. Silicon is the usual cation in the tetrahedral sheets of phyllosilicates. With an ionic radius of 0.26 nm (Table III.1), Si fits snugly in the center of a tetrahedron with four oxygen ions at the corners . Aluminum, and less commonly other small cations, can substitute for Si in tetrahedral sheets.

Octahedra sheet. In phyllosilicate minerals, octahedra are joined together in octahedral sheets (Fig. III-1B and III-1C). The cations in octahedra are in 6-fold coordination. Many cations in the 0.5 to 0.9 nm range (Table III.1) fit into octahedra with six oxygen ions at the corners; for example, Mg, Al, and Fe octahedra.

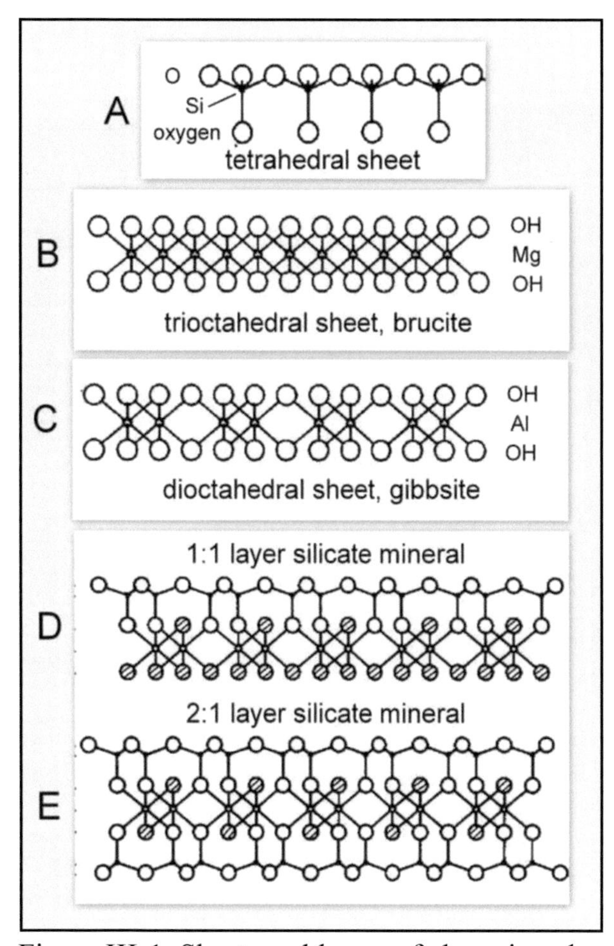

Figure III-1. Sheets and layers of clay minerals. Oxygen ions are represented by open circles and hydroxyl ions by circles containing oblique lines. A. A sheet of silicon tetrahedra. B. A sheet of octahedra with Mg in all octahedral cation positions. C. A sheet of octahedra with Al in two-thirds of the octahedral cation positions. D. A one tetrahedral sheet to one octahedral sheet, or a 1:1 layer silicate, as represented by kaolinite with Al in the octahedral positions. E. A two tetradedral to one octrahedral sheet, or 2:1 layer silicate, as represented by pyrophyllite. Illustrations from Schulze (1989) with permission of the Soil Science Society of America.

Table III.1 Ionic radii of oxygen and cations that occur in the silicate minerals of rocks and soils.

Ion	Coordination Number			
	IV	VI	VIII	XII
O^{2-}	1.38	1.40	1.42	
Si^{4+}	0.26			
Al^{3+}	0.39	0.54		
Ti^{4+}	0.42	0.61		
V^{3+}		0.64		
V^{4+}		0.58		
V^{5+}	0.36	0.54		
Cr^{3+}		0.62		
Cr^{6+}	026	0.44		
Mn^{2+}		0.67, 0.83	0.96	
Mn^{3+}		0.64, 0.58		
Mn^{4+}	0.39	0.53		
Fe^{2+}		0.61, 0.78	0.92	
Fe^{3+}	(0.49)	0.55, 0.64		
Co^{2+}		0.65, 0.74	0.90	
Ni^{2+}		0.69		
Na^{+}			1.18	
Mg^{2+}		0.72	0.89	
K^{+}			1.51	1.64
Ca^{2+}			1.12	

Data from Shannon (1976). Where two IV coordination radii are given for an ion, they are for ions with low and high electron spin configurations. Ions of Fe are too large for good tetrahedral fit, but some tetrahedral Fe(III) has been reported (Borchardt 1989).

Pyroxenes and amphiboles are made of single and double chains of silicon tetrahedra. Spaces between chains are small in orthorhombic pyroxenes (for example, enstatite) and orthorhombic amphiboles (for example anthophyllite), excluding cations larger than Mg. There are larger spaces between the chains of clinopyroxenes and clinoamphiboles, however, that can accommodate Ca and Na in cubic (8-fold) coordination. Large spaces between the tetrahedra of adjacent phyllosilicate layers in micas and vermiculite accommodate potassium (K) in 12-fold coordination.

The layers of common phyllosilicate minerals are constructed of one tetrahedral sheet and one octahedral sheet (1:1 layer silicate minerals, Fig. III-1D) or two tetrahedral sheets and one octahedral sheet (2:1 layer silicate minerals, Fig. III-1E). Chlorite is more complex. No silicate minerals consist of only tetrahedral sheets or only octahedral sheets. Brucite and gibbsite with only octahedral sheets are not silicate minerals.

The main 1:1 layer silicates are kaolinite with Al ions in dioctahedral positions (Fig. III-1D), and serpentine with Mg ions in trioctahedral positions. They contain Si ions in tetrahedral positions. There may be minor amounts of Al substituting for Si to create a net negative layer charge, but kaolinite and serpentine are practically devoid of permanent charge (Table III.2). Most of the negative charges on them are variable charges, created by the dissociation of H^+ from hydroxyl ions at the edges of the octahedral sheets (White and Dixon 2002). There are three main varieties of serpentine: chrysotile, lizardite, and antigorite. The fit of tetrahedral and octahedral sheets is not perfect, requiring adjustments that are different for each variety. In chrysotile, rolling of the sheets into tubes reduces the misfit and creates a fibrous appearance. In lizardite, the layers consist of many small units that crystallize independently and join so as to adjust for the misfit between sheets and create planar layers. In antigorite, small sections of the tetradedral sheets are inverted 180° periodically, creating wavy layers.

The basic 2:1 layer silicates, with only Si in the tetrahedral sheets and without interlayer cations and water molecules, are pyrophyllite with only Al in the octahedral sheet and talc with only Mg in the octahedral sheet. The main 2:1 layer silicates in soils are mica, or illite, vermiculite, and smectite; and also talc in ultramafic soils. Illite, a dioctahedral mica with less Al substituted for Si in tetrahedral sheets than in muscovite, has been reported rarely from ultramafic soils. Vermiculite is not too common as an individual mineral in ultramafic soils, but it is commonly interstratified with chlorite. Smectite is common in ultramafic soils and the dominant clay mineral in many of them. Both vermiculite and smectite have large cation deficits in octahedral sheets and high CECs. Most of the greater negative charge and higher CEC in vermiculite is related to the substitution of Al for Si in the tetrahedral sheets.

Table III.2 Properties of common clay minerals. Most of the common soil clay minerals are layer silicates, except gibbsite. brucite, Fe-oxides, and allophane.

Mineral	Layer[a] Composition	Surface Area	CEC[b]	
			constant	variable
name	tetr:octa	m^2/g	meq/100g	
Goethite	0:0	—	0	< 6
Gibbsite and Brucite	0:1	—	0	< 8
Kaolinite and Serpentine	1:1	10-20	1-2	3-6 or less
Illite	2:1	70-120	15-20	10-15
Vermiculite	2:1	600-800	115-150	< 5
Smectite	2:1	600-800	80-120	< 5
Chlorite[c]	2:1	70-150	10-30	< 20
Palygorskite	chain	140-190	5-30	< 5
Humus[d]	organic	—	variable, 150±	

[a] Ratio of tetrahedral to octahedral sheets in the layers.
[b] Cation-exchange capacity: (1) constant (or "permanent") CEC is independent of soil pH, (2) variable CEC is the difference between the CEC in an alkaline solution and the CEC in a neutral salt solution. Variable charge can be attributed to cation exchange sites that are more negative at higher soil pH than at lower pH. [c] Chlorite includes a diverse group of clay minerals with CEC ranges that may differ greatly from those given here. [d] Soil organic matter, or humus, is added for comparison. Its CEC is dependent on pH, ranging from < 100 meq/100g in very strongly acid soils to 200 or more in alkaline soils.

Vermiculite and smectite are differentiated physically by the expansion of smectite from entry of water between 2:1 layers upon hydration, and lack of expansion in vermiculite. Higher negative charge caused by more substitution of Al for Si in the tetrahedral sheets of vermiculite than in those of smectite, attracts a single interlayer of hydrated cations so strongly that no extra water can enter spaces between 2:1 layers, preventing expansion. Effects of expansive soils with smectite clays in them are evident in Vertisols, which have cracks that close with hydration and open upon seasonal drying.

Smectite occurs as several mineral varieties that are differentiated by the kinds of cations in octahedral positions and substitutions for Si in tetrahedral sheets. Beidellite and

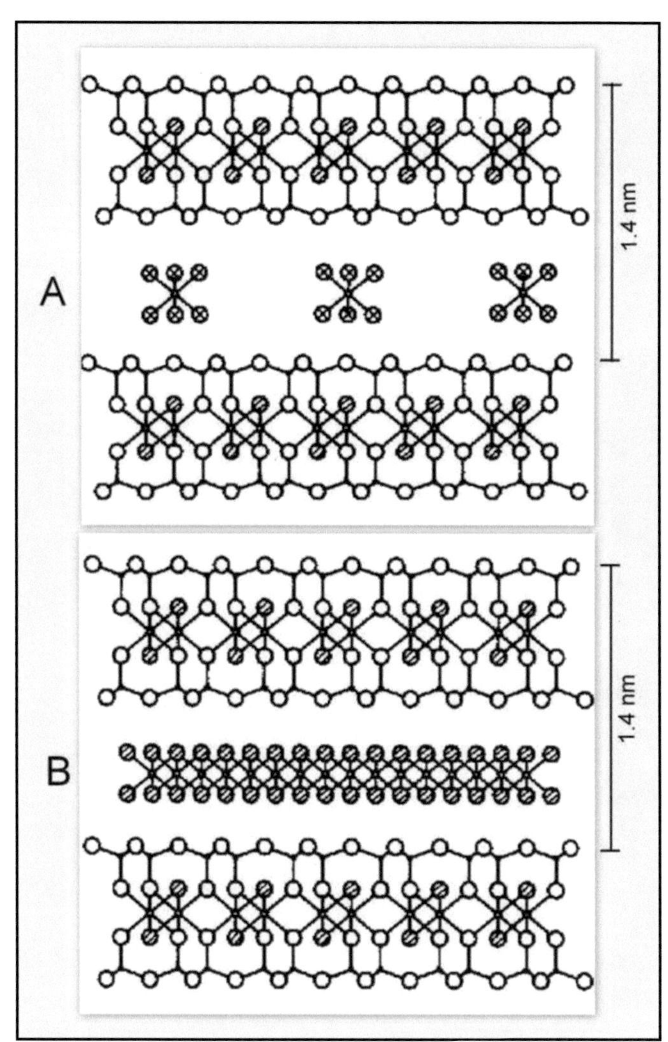

Figure III-2. Phyllosilicate minerals with 2:1 layers having hydrated cations (A) or octahedral sheets (B) between layers. A. A 2:1 layer silicate represented by smectite. B. A 2:1 layer silicate represented by chlorite, which is sometimes called a 2:2 layer silicate to recognize the octahedral hydroxyl sheets between 2:1 layers as structural parts of the mineral. Silicon is represented by dots in the apical and basal planes of each 2:1 layer. Large circles represent oxygen and hydroxyl ions. Small black circles represent cations such as Al, Mg, and Fe in octahedral positions. Illustrations from Schulze (1989) with permission of the Soil Science Society of America.

montmorillonite are common smectites in soils. Nontronite and saponite are smectites that are less common, but more common in ultramafic soils. Both beidellite and montmorillonite have Al in the octahedral sheets. Negative charge is created by substitution of Al for tetrahedral Si in beidellite and Mg for octahedral Al in montmorillonite. The octahedral cations are mainly Fe(III) in nontronite and Mg in saponite, and charge is created by substitution of Al for Si in the tetrahedral sheets. These strictly defined varieties of smectite are generally not found in soils; it is gradations between them that occur in soils (Reid-Soukup and Ulery 2002). A nickeliferous smectite has been called *pimelite.*

Chlorites consist of 2:1 layers with hydroxyl sheets between them (Fig. III-2B). The tetrahedral sheet has a negative charge caused by the substitution of some Al^{3+} for Si^{4+} that is partially balanced by trivalent cations in the interlayer hydroxyl sheet. There is a wide range of chlorites from Mg-clinochlore to Fe-chamosite, with specific gravities from 2.6 to 3.2. The main chlorite in ultramafic soils of North America is clinochlore, confirmed by the flotation of fine sand grains on bromoform (SG=2.89). Experimentally, chlorites weathered to vermiculite or interstratified chlorite/ vermiculite, depending on the mineral compositions (Ross and Kodama 1976). A chlorite with Ni substituted for Mg is called *nimite.*

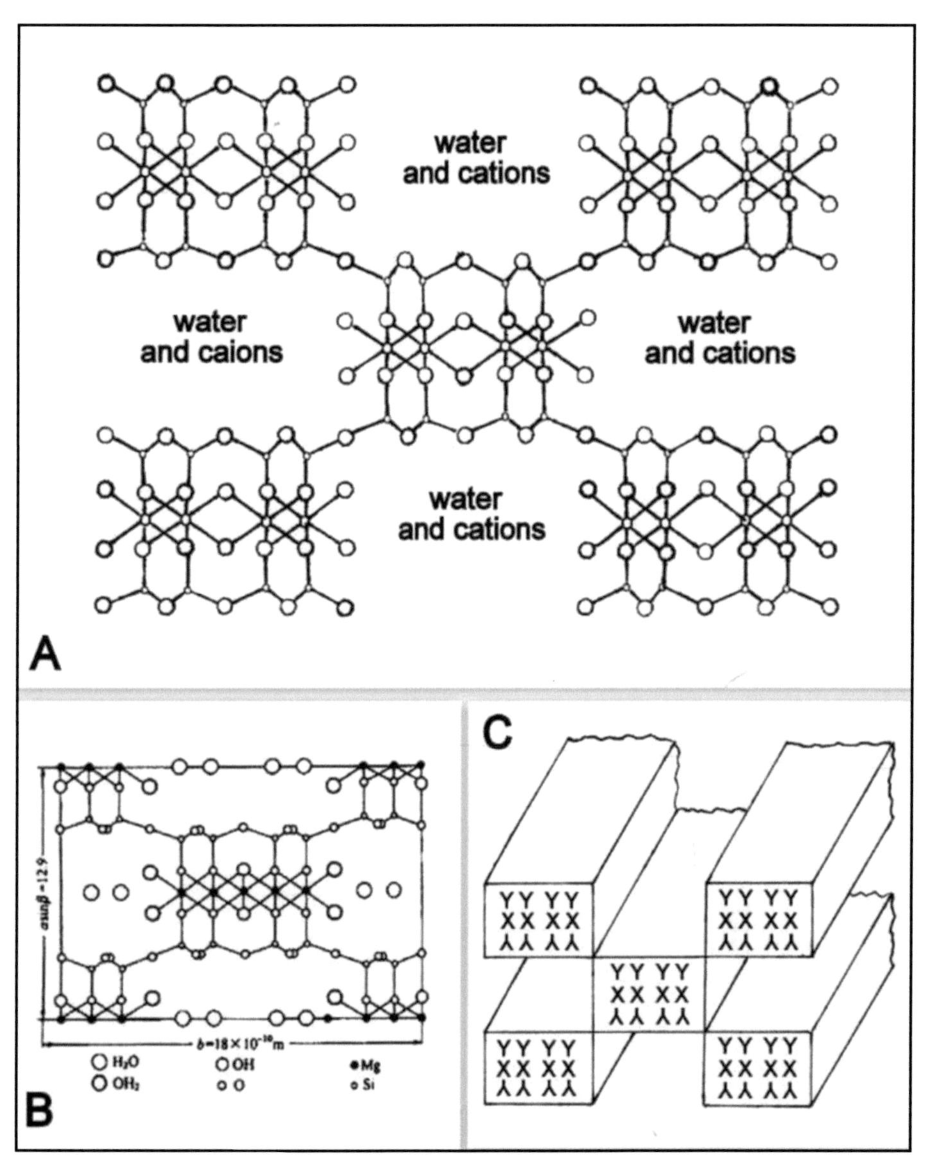

Figure III-3. Palygorskite. A. Structure in a projection perpendicular to the long axis (a) of the ribbons shown in the perspective view (III-3C). B. Cross-section of a crystallographic unit, perpendicular to the long axis. C. A perspective view of palygorskite.

Palygorskite and sepiolite are 2:1 layer silicates in which the layers are ribbons, rather than complete layers (Fig. III-3). They form microscopic fibers with cross-sections that are nearly round in palygorskite and broader in sepiolite (Singer 2002). Sepiolite is orthorhombic and the palygorskite in soils is monoclinic with an axis inclined 105.2 degrees from the long axis of the ribbons.

Models for sepiolite have Mg in all octahedral positions and only Si in tetrahedral positions. In palygorskite there is some substitution of Al for Si in tetrahedral positions, and there is generally Mg, considerable Al, and commonly some Fe(III) in octahedral positions. Some octahedral sites in palygorskite are vacant, not occupied by any cations in them.

Palygorskite is common in soils of warm and hot arid climates. Sepiolite is commonly found in saline sediments or sedimentary environments. A nickeliferous sepiolite has been called *falcondoite*.

IIIA2 Weathering and clay minerals in ultramafic soils

The main minerals in dunite and peridotite are forsterite (Mg-olivine), orthopyroxenes, clinopyroxenes, and chromite. Metamorphism produces serpentine, brucite, and magnetite, and commonly some chlorite. Talc is less common in ultramafic rocks.

Weathering of olivine is complex (Delvigne et al. 1979). The first product of weathering in olivine may be saponitic (Mg-rich) smectite that becomes more nontronitic (Fe-rich) as Mg is leached out and Fe is retained or added from the weathering of associated minerals. Extra Fe forms Fe-oxides and oxyhydroxides.

Serpentine is less readily weathered than olivine and pyroxenes. Brucite is so easily weathered that it is never reported in soils investigations. Chromite and magnetite are very resistant to weathering. The chlorite from metamorphosed ultramafic rocks generally has predominantly Mg in octahedral positions. In the initial stages of weathering, chlorite might be interstratified with smectite or vermiculite.

The clay mineral compositions in relatively old ultramafic soils are dependent on climate. Iron oxides accumulate in humid soils, smectites in summer-dry soils, and palygorskite in warm or hot arid soils. Some unique minerals are found in Ni-laterites (Appendix IV)..

B. Investigations of Clay Mineralogy in Ultramafic Soils

Clay mineralogy in ultramafic soils has been investigated in many locations on all continents other than Antarctica. The results are similar to those for ultramafic soils in North America and the Caribbean area. Most of the investigations have been for specific sites or landscapes.

A summary of results from investigations throughout Scotland indicates some common trends of clay transformation in ultramafic soil development. In Scotland the main ultramafic soils are loamy Alfisols, commonly with lizardite and minor chrysotile (Wilson and Berrow 1978). Serpentinite weathers to produce soils with serpentine, chlorite, interstratified chlorite/vermiculite, smectite, and minor talc. Smectite ("ferruginous saponite") is generally dominant in subsoils and diminishes upward in the well drained soils.

Although the trends of clay mineral composition in the development of many ultramafic soils may be similar to that in Scotland, the details may differ greatly. There are many different patterns of ultramafic clay mineral development in soils of eastern and western North America that are discussed in this section (IIIB). The development of very old soils in which large amounts of Fe oxides accumulate is presented in Appendix IV.

IIIB1 Eastern North America.

In Québec, De Kimpe and Zizka (1973) found that serpentinized dunite containing serpentine, brucite, and magnetite weathered by way of interstratified minerals to produce chlorite and smectite in an Inceptisol. Also, De Kimpe et al. (1987) found that all primary minerals in pyroxenites at Mont Saint-Bruno weathered to produce smectites.

On the Appalachian piedmont northwest of Chesapeake Bay, Rabenhorst et al. (1982) studied four serpentine Alfisols (Hapludalfs) with moderate to minor components of loess in them. The soils were strongly to very strongly acid in A horizons, with acidity decreasing with depth to slightly acid or neutral in B and C horizons. Serpentine minerals were common to plentiful in the coarse clay and silt fractions, but practically absent from the fine clay <0.2 μm. Smectites dominated the clay fractions in one soil and chlorite and vermiculite were dominant in the others. Most of the vermiculite was interstratified with chlorite.

In the Blue Ridge Mountains, North Carolina, Ogg and Smith (1993) studied two deep Alfisols (Hapludalfs) derived from altered peridotites containing tremolite, or actinolite, and chlorite, with no olivine. The main clay minerals were interstratified chlorite/vermiculite, kaolinite, and talc, with Fe-oxides in goethite. An Inceptisol developed from talc-chlorite schist was described above Buck Creek in the Nantahala National Forest (Chapter 7). The practically unweathered schist in the C horizon of the Inceptisol contained mostly chlorite (apparently clinochlore) and some talc, which was absent from the B horizon, leaving only chlorite in the clay fraction of the cambic horizon.

On the Llanos uplift of central Texas, soils on serpentinized peridotite or serpentinite are Mollisols. Maoui (1966) found that the fine clay in the Renick soils was mainly smectite and the coarse clay was predominantly serpentine and chlorite, with the serpentine more concentrated in the C horizon and chlorite in the upper horizons. Some quartz, clinoamphibole of the tremolite-actinolite series, and talc were found and may be from schists that are closely associated with the serpentinite, although the clinoamphibole and talc are commonly formed by alteration of the same peridotite from which the serpentine was produced by serpentinization.

IIIB2 Northwestern North America.

In cold climates of southwest British Columbia, Bulmer and Lavkulich (1994) found that the common clay minerals in ultramafic soils were serpentine, chlorite, and talc, and minor smectite. Also, in the wettest of three areas where they sampled the soils, they found an interstratified clay that they designated vermiculite/smectite.

In Southeast Alaska (Alexander et al. 1994b), a silica-cemented hardpan beneath a Spodosol in mostly nonserpentine material was in glacial till containing olivine, clinopyroxenes, serpentine, magnetite, and a Si-cemented hardpan. Only serpentine was detected in the clay fraction from the hardpan.

IIIB3 Central Pacific Cordillera

In the California region of summer-dry climates, kaolinite is very common, but absent from ultramafic soils (Graham and O'Geen 2010). Smectite is a common mineral in the ultramafic soils. Istok and Harward (1982) found that cool (mesic STRs) very well drained ultramafic soils were less favorable than wetter soils for the development of smectite in them. In warm soils, Wildman et al. (1968) identified ferruginous smectites in well drained soils, whereas Senkayi (1977) found magnesium smectites in lower topographic positions where silica and Mg can accumulate in the sediments and soils. McGahan et al. (2008) investigated the distributions of Ca-bearing minerals in some soils of oceanic terranes in the California Coast Ranges.

Graham et al. (1990) investigated the mineralogy of a moderately deep, cool (mesic STR) serpentine Mollisol (Argixeroll) with A, Bt, and Cr horizons in an open Jeffrey pine/buckbrush (*Ceanothus cuneatus*)–whiteleaf manzanita (*A. viscida*)/Idaho fescue forest in the Klamath Mountains. The parent rock was a highly serpentinized peridotite with a small amount of enstatite remaining in the serpentinite. Fine sand from the Cr horizon was mostly serpentine with small amounts of magnetite, chlorite, and enstatite. Serpentine decreased upward toward the A horizon and magnetite increased to become a dominant mineral in the fine sand fraction of the A horizon. Serpentine and chlorite were common in the clay fractions and smectite appeared below the A horizon and increased in abundance downward to the Cr horizon.

In a humid climate, with annual precipitation about 3000 mm, just west of the Klamath Mountains, Burt et al. (2001) found serpentine to be the dominant clay mineral in a cool Inceptisol and a cold Alfisol with serpentinized peridotite parent materials. There was some chlorite and vermiculite in the Inceptisol and minor vermiculite in the surface and minor smectite in the subsoil of the Alfisol.

Lee et al. (2003) investigated a toposequence of cool, well drained to wet ultramafic soils in colluvium over serpentinized peridotite in the Klamath Mountains. Serpentine and chlorite were the main minerals in the colluvium, with minor talc, amphibole, and quartz. Serpentine and some chlorite remained in the soils, but low-charge smectite was the main mineral in the clay fractions. By loss of interlayer hydroxyl interlayer sheets, chlorite was transformed to yield interstratified chlorite/vermiculite and vermiculite.

Table III.3 Citrate-dithionite extractable Fe in subsoils (30-45 cm depth) of cool (mesic STR) ultramafic soils along a transect across the Klamath Mountains from the arid interior to the humid Pacific coast at about 42°N latitude. The soil at transect site 20 is on a very old surface.

Trans. Site No.	km to Pacific Coast	MAP[a] mm	Soil Class. USDA Soil Taxon.	Fine Sand Minerals[b] H:L:M	pH DW	Fe_d (g/kg)	Vegetative Cover
02	136	400	Argixeroll	20:76:4	6.4	15	steppe
09	95	1100	Haploxeralf	88:11:1	6.3	46	open forest
15	33	2100	Hapludalf	7:84:9	6.2	47	forest/shrubs
18	13	2800	Hapludalf	50:8:42	6.1	153	forest/shrubs
21	1	2000	Argiudoll	21:63:16	6.2	77	dense forest
20	13	3000	Kanhapludalf	18:41:41	6.3	189	secondary forest

[a] MAP, mean annual precipitation. [b] H:L:M proportions are the relative weights of heavy nonmagnetic (H), light (L), and heavy magnetic (M) grains in the fine sand (0.125-0.25 mm) separates by bromoform (SG=2.89) flotation and a hand magnet.

Cool (mesic STRs) soils from the dry inland side to the wet coastal side of the Klamath Mountains all have serpentine, and chlorite, which are inherited from the ultramafic rocks, but these minerals are nearly gone from the more completely weathered soil at site 20. Smectite is dominant in the drier inland soils and goethite is the dominant clay mineral in the more moist soils nearer the coast (Alexander 2014a, Chapter 9, Fig. III-4).

Lacking heat treatments, it was not possible to separate vermiculite and interstratified chlorite/vermiculite from chlorite, but based on the reports of Rabenhorst et al. (1982), Lee et al. (2003), and Alexander et al. (2007b), some interstratified chlorite/vermiculite is likely to be present and possibly minor amounts of vermiculite. Fine sand in the Alfisol at site 09 is predominantly pyroxenes, with some olivine, and the clay fraction is dominated by smectite more than any other of the soils.

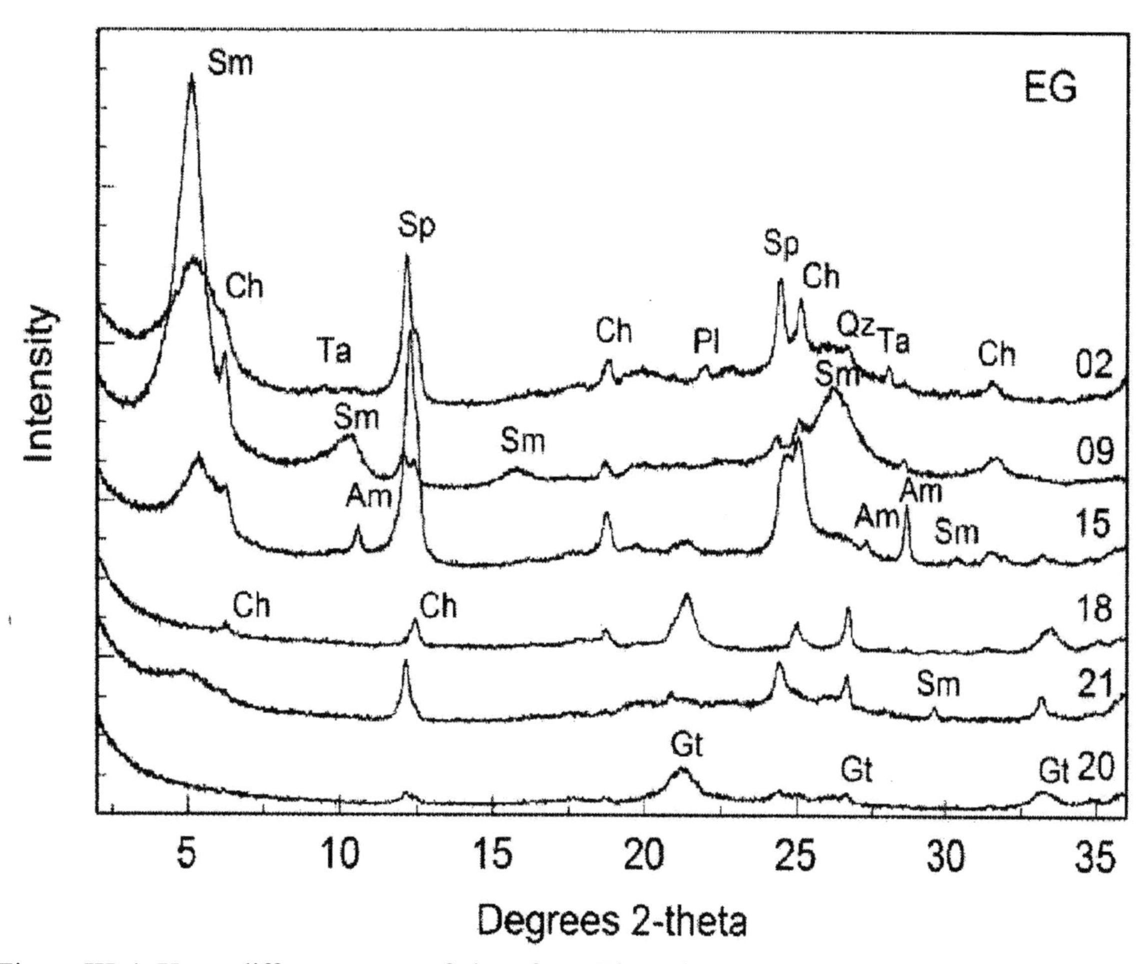

Figure III-4. X-ray diffractograms of clays from Klamath Mountain soils (Table III-3). Subsoil samples with ethylene glycol applications. Am, amphibole, (tremolite-actinolite); Ch, chlorite; Gt, goethite; Pl, plagioclase; Qz, quartz; Sm, smectite; Sp, serpentinite; Ta, talc.

IIIB4 Arid ultramafic soils, Baja California

With minor exceptions in the Great Basin, all of the arid ultramafic soils of North American are in Baja California. Most of the arid ultramafic soils are in mountainous terrain and lack appreciable morphological development. The older ultramafic soils on a sequence of arid terraces were found to have moderately alkaline, reddish brown to yellowish red argillic horizons and thick duripans (Chapter 10, sections 10C2 and 10D). Clay minerals in the youngest soil, an Entisol (Torrifluvent, 1AC in Figure III-5), were serpentine and minor chlorite, smectite, and palygorskite, with some allochthonous (aeolian) quartz. The same minerals were present in older soils (argillic Aridisols), with much more palygorskite (Figure III-5). Palygorskite was dominant in the oldest soil, an argillic duric Aridisol (Duric Petroargid), and there was no smectite in it.

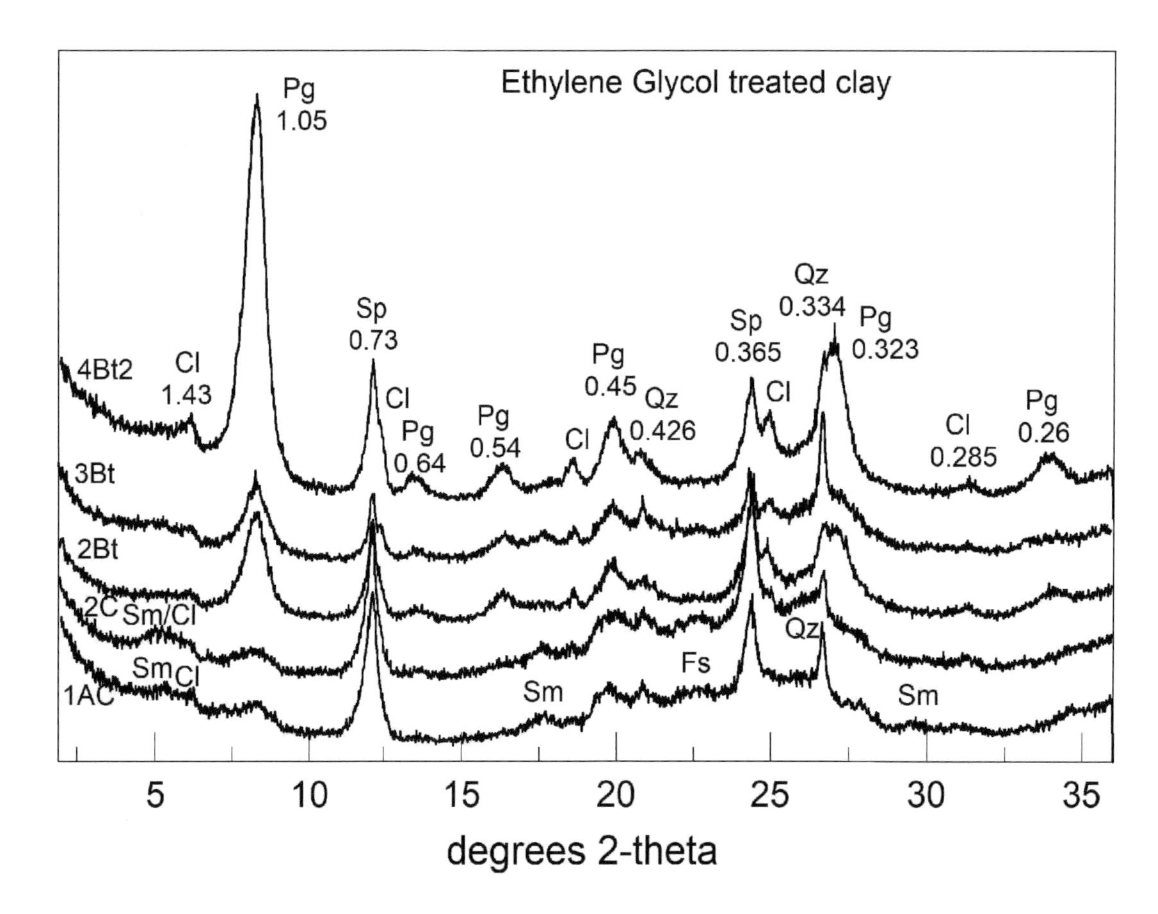

Figure III-5. X-ray diffractograms of clays <4 micrometers from Baja California magnesite mine soils (Chapter 10, section D). From the bottom to the upper line, the samples are from pedons 1 to 4; AC, C. Bt, and Bt2 horizons. The minerals are Cl, chlorite; Fs, feldspars; Pg, palygorskite; Qz, quartz; Sm, smectite; and Sp, serpentine.

Appendix IV

Extreme Weathering – Oxisols and Ni-laterite

Intensive weathering in hot climates for millions of years is generally necessary to produce Oxisols and Ni-laterites. In North America, these conditions exist only in the Caribbean area. There are small areas of Oxisols in California, however, that were formed in sediments of the Ione formation during the Paleogene, and Ni-laterites are present on a Miocene peneplain in the Klamath Mountains of California and Oregon.

. The residue from extreme tropical weathering and leaching is essentially Al and Fe oxides and oxyhydroxides. Most soil parent materials contain more Al than Fe and the result of extreme weathering and leaching is usually bauxite.

Bauxite – Aluminum ore deposits lacking iron.
Most rocks contain substantial aluminum, which is relatively immobile in soils. Much of the Al is retained as rocks and sediments weather and other elements are removed by leaching. The result after millions of years of Al accumulation is bauxite. Paleogene bauxite is mainly gibbsite, $Al(OH)_3$, and Mesozoic bauxite is mainly boehmite, (AlOOH). Bauxites are commercial sources of aluminum. Ultramafic rocks have insufficient Al for the formation of bauxite.

Very old soils developed from parent materials with ultramafic compositions are very different from soils with more aluminum. Fe-oxides and oxihydroxides are concentrated in them. Tropical soils with high Fe-oxide concentrations harden upon drying. For many centuries, reddish tropical soils in southern India had been cut into blocks that were dried in the sun to produce building blocks. This process was observed by Buchanan (1807) who named the indurated blocks of soil laterite (L. *later*, brick). Soils with high Fe-oxide contents became known as *laterite* soils, even though the moist *laterite* soils hardened only when exposed to the atmosphere for long periods of time. Fe-cemented crusts that form in seasonally dry soils are called ferricrete.

A. Ni-laterite

The term *laterite* has lost favor among pedologists, but Ni-laterite is a term that is in common use by geologists for Fe-oxide deposits in which Ni and Co are concentrated along with the iron in intensively weathered regolith. Because of their economic importance as sources of the transition elements from Cr to Fe, Co, and Ni, Ni-laterites have been investigated comprehensively, particularly in Australia (Brand et al. 1998), Brazil, and New Caledonia (Trescases 1975).

Ni-bearing lateritic soils were mapped on ultramafic rocks in Cuba early in the 20th century (Bennett and Allison 1928), but economic geologists had already been investigating Ni-laterite deposits for decades (Beck and Weed 1905). The Ni-laterite deposits in New Caledonia were discovered and described by Garnier in 1867. Garnierite, which is a mixture of nickeliferous varieties of serpentine, chlorite, talc, sepiolite and/or smectite, has been named after him. Some of the Ni varieties of the phyllosilicates have been called népouite, nimite, pimelite, willemseite, and falcondoite.

A typical Ni-laterite profile (Fig. IV-1) generalized from descriptions and data of Rice (1957), Hotz (1964), and Trescases (1997) shows (1) bedrock, (2) weathered rock, (3) saprolite with boxwork of silica (opal, chalcedony, or quartz) and garnierite, (4) a transition zone with black mottles or coatings, (5) a yellowish subsoil dominated by goethite, (6) a reddish soil with goethite plus hematite and maghemite, and (7) a thin A-horizon with several percent organic matter. In seasonally dry climates, ferricrete can occur at the ground surface of Ni-laterite profiles. Clay minerals in the lower saprolite are serpentine, chlorite, smectite (Mg-smectite), and if there is much pyroxene in the parent rock, nickeliferous talc. Upward through the weathered saprolite, there is less serpentine and chlorite and more smectite (Fe-smectite). The Ni concentrations are commonly about 2% in the upper saprolite, or higher if much Ni-talc is present, and about 1% in the limonite above the saprolite. Cobalt is concentrated along with Mn-minerals in the black deposits called absolane that are in the transition from the limonite to the saprolite. The absolane generally contains about 0.1 to 0.2 percent Co.

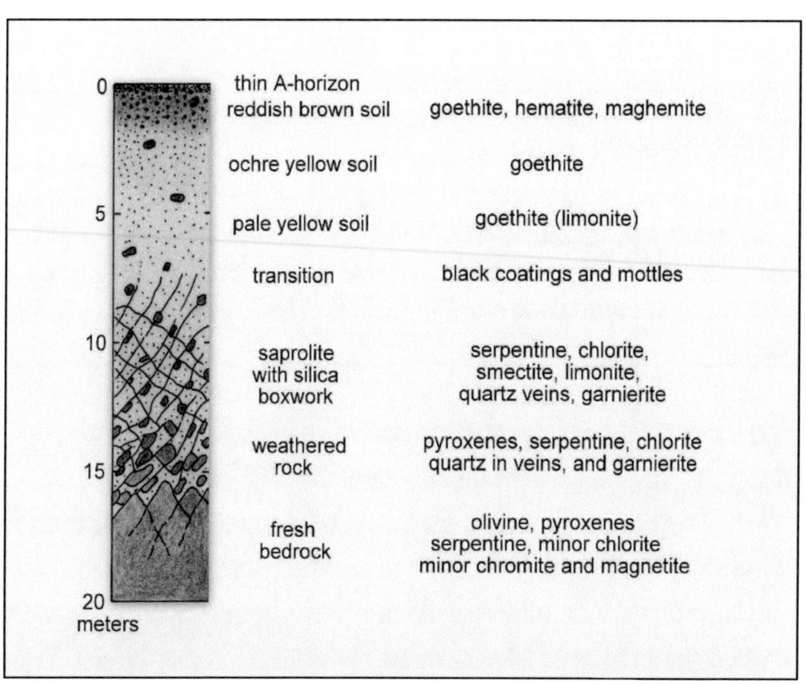

Figure IV-1. A schematic representation of Ni-laterite based on descriptions and data from Rice (1957), Hotz (1964), and Trescases (1997). The black deposits in the transition zone are called absolane.

Nickel in Ni-laterites that are commercial sources of Ni are commonly concentrated to 1.0 to 1.6% Ni in limonite ore (about 2% in goethite) and to a mean of 1.44% in "hydrous Mg-silicate" ore of saprolite (Butt and Cluzel 2013). There are numerous varieties of garnierite. The best ores containing garnierite have 2 to 5%, or more, Ni (Butt and Cluzel 2013).

B. Oxisols

Ultramafic Oxisols and Ni-laterites are found in Cuba, Guatemala (Valls 2006), Hispaniola, and Puerto Rico (Lewis et al. 2006). The Nipe soil is a well known representative of Caribbean ultramafic Oxisols. It was mapped by Bennet and Allison (1928) in the Nipe area of Cuba and later in Puerto Rico (Roberts 1936). It was thoroughly described and characterized in a 1982 issue of Geoderma. The morphology of the Nipe soil and the elemental chemistry reported to depths >150 cm (Beinroth 1982) are comparable for samples from Cuba (Bennett and Allison 1928) and from two locations in Puerto Rico (Roberts 1936, Sivarajasingham 1961).

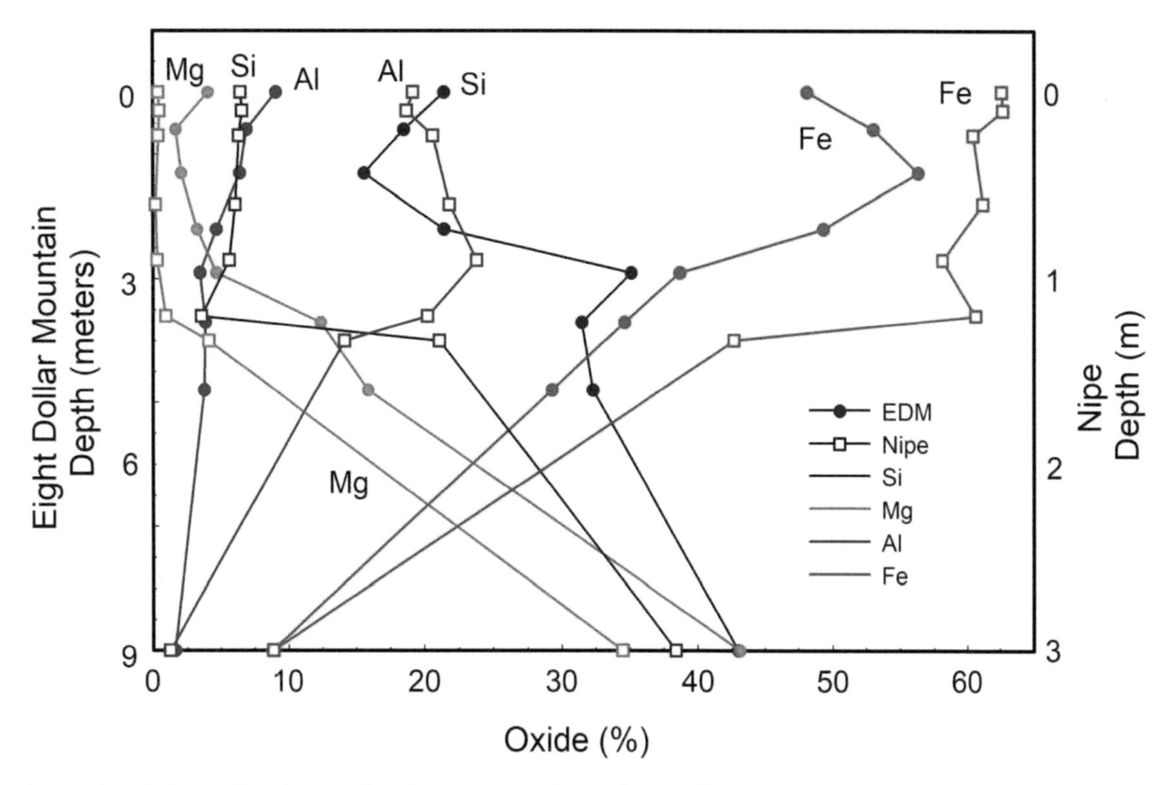

Figure IV-2. Distributions of major cations from the surface down into bedrock at 9 meters in a Ni-laterite on Eight Dollar Mountain (EDM), Oregon (Hotz 1964), and down to bedrock below 1.5 meters in the Nipe soil of Puerto Rico (Sivarajasingham 1961).

Silicon and Mg are the dominant elements in ultramafic rocks. Ultramafic rocks have low Al contents and relatively high Fe contents, compared to most of the more common soil parent materials. With intense weathering and leaching much of the Si and nearly all of the Mg are removed from ultramafic soils and Fe becomes a dominant cation. Aluminum is relatively immobile in soils and accumulates in soils as much Si and most of the Mg is removed by weathering and leaching.

A comparison of the major chemical element distributions in a Nipe soil and in the Ni-laterite on Eight Dollar Mountain in Oregon (Fig. IV-2) shows similar trends for Fe and Mg in both, but the loss of Si is much greater from the Nipe soil. With the greater loss of Si, Al has become a more prominent cation, with appreciable amounts of kaolinite and gibbsite in the Nipe soils (Fox 1982, Jones et al. 1982). Kaolinite and gibbsite are absent from ultramafic soils in California and Oregon (Foose 1992). The increased loss of Si from more intense weathering and leaching in warmer climates, and retention of Al, is a trend recognized in Brazil, where there are ferruginous Ni-laterites in tropical and subtropical regions (Barros de Oliveira et al. 1992). There was enough Al in the Ni-laterite soils of the tropical areas in Brazil for the formation of kaolinite and gibbsite in them.

Appendix V

Soil Sampling and Analyses

Pedon sampling, sample preparation, and estimation of coarse fragment volume (%)
Volumes of stones (cobbles and coarse gravel, 30-600 mm fraction) were estimated visually in soils when they were sampled (Vs). Samples of soil <30 mm were air-dried and sieved to obtain fine-earth (particles <2 mm) for laboratory analyses. The volume of 2-30 mm gravel retained on the sieve was estimated visually, or weighed and converted to volume (Vg) by assuming that the density of the gravel was twice that of the bulk density of the soil. The field, or total, volume of "coarse fragments" (Vf) was computed by the field estimate of stones plus the lab estimate of 2-30 mm gravel times 100 minus the field estimate of stones: Vf = Vs + Vg(100-Vs). Note: all of these volumes are percentages.

Hygroscopic water
Air-dry soils were dried in a microwave oven to determine ratios of air-dry:oven-dry weights for converting laboratory measurements to oven-dry bases.

Soil acidity (pH) by electronic meter with a glass electrode
Soil pH was recorded in a 1:1 distilled water:soil suspension (DW), and in some cases, in a 2:1 molar KCl solution:water suspension (designated KCl).

Soil acidity-alkalinity (pH scale) in indicator solution

Indicator solutions, pH from LaMotte color charts

- bromocresol green, BCG (pH 3.8 to 5.0)
- chlorophenol red, CPR (pH 5.2 to 6.2)
- bromothymol blue, BTB (pH 6.2 to 7.4)
- phenol red, PHR (pH 6.8 to 7.8)
- cresol red, CRR (pH 7.4 to 8.6)
- thymol blue, THB (pH 8.2 to 9.0)

Aqua Regia digestion, Inspectorate America Corporation, Sparks NV

(1) a small sample pulverized for at least 85% of it to pass a 200 mesh (75 µm) screen
(2) hot modified aqua regia digestion (an aggressive, but not entirely complete digestion)
(3) evaluation of chemical element concentrations by ICP

"Free" iron and other first transition elements, citrate-dithionite extraction

Iron, Mn, and Cr were reduced by Na-dithionite in 0.3 molar Na-citrate in an erlenmeyer flask, with occasional shaking and additions of slightly more Na-dithionite each day for two or three days before filtering to obtain a solution for analyses of Fe, Mn. Cr, Co, and Ni by AA spectrometry. Alternatively, citrate-dithionite extracted iron (Fe_d) was ascertained by adding o-phenanthroline to develop a color and recording the relative amount of transmitted light, following a procedure of Holmgren (1967).

Exchangeable cations

Cations were extracted be leaching with KCl or NH_4Cl and ascertained by AA spectrometry. Alternatively, Ca and Mg were titrated with EDTA (Heald 1965).

Exchange acidity, pH 8.2

Soils were leached with pH 8.2 BaCl-triethanolamine solution and the leachate was titrated with HCl to a pink end point (pH 5.1) of a mixed bromocresol green and methyl-red indicator (Peech 1965). The exchange acidity is the difference between the amount of acid used to titrate the leaching solution and the amount of acid used to titrate the leachate.

Exchange acidity, pH 7.0 (*acid soils, pH <7.0) only*)

The method is practically the same as the pH 8.2 exchange acidity method, except that the buffer solution is para-nitrophenol, rather than triethanolamine. Soils were leached with pH 7.0 KCl-PNP (*p*-nitrophenol) solution and back titrated with HCl (Alexander 2010).

> Make a buffer solution by dissolving 7 g of *p*-nitrophenol (13.9 g/mol) in a liter of 0.5 M KCl or NH_4Cl and adjusting to pH 7.0 by adding KOH or NaOH. The *p*-nitrophenol will not dissolve completely until some of the hydroxide has been added to raise the pH. Make a replacement solution by adding 10 ml of buffer solution per liter to 0.5 M KCl or NH_4Cl. Pipette exactly 25 ml of the buffer solution into an Erlenmeyer flask containing 5.00 g of soil. Shake the flask occasionally, and after 2 hours, wash through a filter with three 20-ml portions of replacement solution. Add 10 drops of mixed bromocresol green and methyl-red indicator solution to the leachate and titrate with 0.5 M HCl to a red endpoint.

Extractable acidity, ambient pH

Soils were leached with molar KCl solution, and the leachate was titrated to a pink phenolphthalein end point with HCl (Yuan 1959).

Note. The Soil Survey Laboratory Methods Manual (Burt 2004) of the National Soil Survey Laboratory is a good source of information about laboratory procedures.

Glossary

words and phrases as they are applied in geoecology

absolane - A black manganese oxyhydroxide in which Co is concentrated in Ni-laterites.
aeolian - see *eolian*
algae - Immobile organisms that live in water and fix carbon by photosynthesis. Some are eubacteria, some are protoctists, and some may be primitive plants.
allochthonous - Something that has been formed in one place and is found in another; it is something that is foreign, or nonnative; sometimes referred to as *exotic*.
alluvial, alluvium - Sediment deposited from suspension in moving water.
anion - A negatively charged ion.
autochthonous - Produced or formed where it occurs presently.
available water (soil) - Water available to plants through their roots.
available-water capacity (soil) - The amount of soil water that a soil wetted to field capacity (approximately -10 kPa, or 0.1 bar suction) will release to plants before common cultivated plants wilt (about -1.5 MPa, or 15 bar suction).
barren (ecology) - An area with scant vegetative cover, or none at all.
bog - Variously defined wet land or swamp. A bog might be differentiated from a fen by being acid. Bogs commonly contains peat, and sphagnum, and ericacious shrubs are common.
cation - A positively charged ion of a chemical element..
cation exchange - The interchange of positively charged ions among solids or between solids and solutions surrounding the solids.
cation-exchange capacity - The capacity of solids (for example clay minerals and soil organic matter) to retain cations on sites from which they can be displaced by other cations that then occupy the sites. It is commonly designated as CEC.
chalcophile - Copper, zinc, and other elements that have an affinity for sulfur. Chalcophile elements are common in sulfides.
chamaephyte - A small shrub with buds close to the ground where the deleterious effects of winter winds are minimized.
chronosequence - A sequence of related things (for example, soils) that are differentiated by their ages.
clay - Fine grained (particles < 2 μm) inorganic material in sediments or soils.
clay mineral - Any mineral, generally a layer silicate, that occurs in very small particles that have nominal diameters < 2 μm.
colluvial, colluvium - A loose, heterogeneous or incoherent mass of soil material or rock fragments deposited mainly by mass wasting, commonly on a footslope at the base of a steep slope.
concretion (soil) - A discrete, more or less spherical or concentrically zoned mass of cemented soil or chemical deposit, such concentric layers of calcite.
consolidated material (geology) - A coherent mass of material that has been compacted or indurated.
craton - The core, or more stable part, of a continent that has not been engaged in accretionary processes for a very long time. Its extent is somewhat arbitrary; for example, the Laurentian craton may, or may not, include the Grenville terranes from Newfoundland

to northern Mexico, depending on a persons perception. The North American craton more definitely includes the Grenville terranes, and areas where the Laurentian craton has been covered by Phanerozoic sedimentary deposits. Cratons generally have magmatic and tectonic activity on them, such as the Rocky Mountains, and marine sedimentary cover, such as on the interior of North America.

crust - The outermost layer of Earth, above the mantle.

diamicton - A massive conglomeration of unsorted coarser and finer materials such as till.

ecosystem - A community of living organisms and the environment around them.

edaphology - A study of influences of soil on plant growth and productivity.

eluviation - The removal of material in solution or suspension from a layer of soil.

endemic - A class of plants, or other living organisms, that is restricted to limited geographic localities or occurs only on specific substrates, such as ultramafic rock or serpentine soil.

eolian - Material transported by wind.

epipedon - The uppermost layer of soil in a pedon.

eukaryote - A cellular organism with a nucleus that is bound in a membrane.

evaporation - The conversion of liquid to gas–for example, water to vapor.

evapotranspiration - The combination of water evaporated from the surface of the earth and water transpired by plants.

fen - variously defined wet land or swamp. A fen might be differentiated from a bog by being neutral or alkaline. A fen commonly contains muck, and grasses and sedges may be the main vegetative cover.

ferruginous (regolith) - Reddish soil, or regolith, with much ferric iron.

footslope - A comparatively gentle slope from the base of a steep mountain or hill.

garnierite - A green amalgamation of Mg-Ni phyllosilicates. It is composed of nickeliferous lizardite (or nepouite) and any one or all of talc, smectite, chlorite, and/or sepiolite. The shade of green, or green chroma, increases with Ni content (Brindley and Huang 1973).

geoecology - The study of living organisms in their natural environments that include rock, soil, water, and air.

geoecosystem - An ecosystem with the *geo* prefix added to assure the rock, soil, groundwater, and atmosphere are included, because nonliving parts of ecosystems are commonly neglected.

glacial drift - A general term for all kinds of materials deposited from a glacier by different processes.

glacial till - Unsorted glacial drift.

glacier - A large perennial mass of ice that moves slowly under stresses from its own weight.

graminoid - A grass or a similar plant with long narrow leaves; the Poaceae (grasses), Cyperaceae (sedges), and Juncaceae (rushes) are common families of graminoid plants.

humus - Organic matter in soils that has decayed sufficiently that its origin is no longer evident by cursory examination.

hydration - The addition of water to an organic or inorganic solid.

hygroscopic water - Water retained on air-dry surfaces.

hyperaccumulation (plants). Exceptionally large accumulation of a chemical element in a plant. For Ni and most other heavy elements ($z > 20$), concentrations greater than 1000 µg/g (dry).

illuviation (soil) - The deposition of material that has been moved downward in solution or suspension from one soil layer to another.

incompatable elements - Elements with large ions, such as K^+ and Ca^{2+}, or elements with high charge, such as pentavalent phosphorous (P^{5+}), that are excluded from olivine and orthopyroxenes. Very large K^+ is excluded from clinopyroxenes, also. The trivalent ions of Al and Cr may be incompatible in olivine and pyroxenes, but they are common in the spinel and garnet minerals of ultramafic rocks.

induration (soils) - Hardening by cementation or by other means.

infiltration (soil) - Passage of water from above ground into a soil.

ionic potential - An indication of the polarizing power of an ion, a cation in particular. It is the ratio of the electrical charge of an ion to its ionic radius.

ionization potential - The energy required to remove an electron from an atom or a molecule.

laterite -Soil or regolith infused with iron oxides that harden irreversibly upon drying.

lateritic soil - Highly weathered soil in which aluminum and iron are concentrated by loss of other elements. Iron is highly concentrated in lateritic soils derived from the weathering and leaching of ultramafic parent materials.

leaching (soil) - Transport of material within a soil, or out of a soil, as a solute or suspension in water.

ligand - An anion or a neutral molecule that combines with metallic cations to form complexes.

mafic (igneous) rock - An igneous rock composed chiefly of one or more dark ferromagnesian, silicate minerals.

magma - Naturally occurring molten material within Earth that upon solidification underground or extruded to solidify above ground forms igneous rocks.

mantle - The interior of Earth, below the crust and above the core. By volume, Earth is 84% mantle.

mass wasting - Down slope movement of rock or other detritus by gravity, either dry or in a fluid mass, rather than suspended in flowing water. Rock fall, creep, landslide, and debris flow are some the kinds of mass wasting.

mélange - A heterogenous mixture of rock materials,

metamorphism - The mineralogical and structural adjustment of rocks to altered physical and chemical conditions at depths where the temperature and/or pressure are substantially greater than at the ground level.

metasomatism - The process of practically simultaneous capillary solution and deposition by which a new mineral of partly or wholly different composition may grow in the body of an old mineral or mineral aggregate.

meteoric water - Water from precipitation of moisture in the atmosphere, rather than water released from magma or from the pore spaces of buried rocks.

mineral - A naturally occurring solid in which the atoms have a specific arrangement.

nodule (soil) - A discrete, spherical to irregularly shaped mass of indurated soil or weathered rock within a soil.

obduction - The pushing of ocean crust onto a continent along a continental margin.

octahedron (mineralogy) - A three dimensional crystal structure with eight triangular sides.

olistostrome - A chaotic mix of heterogenous materials that accumulated as a semi-fluid body by submarine gravity sliding or slumping of unconsolidated regolith and rocks.
ophiolite - A sequence of rocks from peridotite through gabbro and sheeted dikes to basalt that is produced on ocean floors and becomes transplanted onto a continent. Most ocean crust is subducted into the mantle; only that which is preserved by emplacement on continents is called ophiolite.
orogen, or orogene - A region where there has been an orogeny.
orogeny - Alteration of a portion of Earth's crust by disruptive forces that cause folding, faulting, and thrusting.
oxyanion - A complex anion consisting of a cation and oxygen; for example, the trivalent oxide of phosphorous (PO_4^{3-}).
oxyhydroxide - A compound that contains both oxygen and hydroxyl ions; for example goethite, FeOOH.
parautochthonous - Tectonically intermediate between autochthonous and allochthonous.
parent material (soil) - Rock or unconsolidated material from which a soil has developed.
pedology - The study of the upper part of regolith, where changes in it are effected by meteoric water, the atmosphere, and biological activity.
pedon - A three-dimensional body of soil with a ground surface of 1 to 10 m^2 and deep enough to reveal soil down to bedrock or other material from which the soil developed or rests upon.
peneplain - A nearly featureless, flat to gently undulating land surface of considerable area which presumably has been produced by subaerial erosion. The original peneplain surface may have been largely dissected subsequently, as in the Klamath Mountains where there are few of the flatter land surfaces remaining, but there are many concordant mountain summits that represent a former peneplain.
petrography - A discipline focused on the description and classification of rocks.
photosynthesis - The production of carbohydrates by plants utilizing energy from solar radiation. It is a process in which carbon from carbon-dioxide is fixed as organic carbon.
phyllosilicate - A layered mineral composed of sheets of silicon tetrahedra combined with a sheet of octahedra containing aluminum or other cations. Examples are mica, serpentine, talc, and clay minerals such as smectite and kaolinite.
physiognomy (vegetation) - A physical description of vegetation, without regard to plant taxonomy.
physiography (geology) - The physical description of landscapes.
plinthite - Friable soil in reddish mottles or layers that hardens irreversibly upon drying..
precipitation (meteorology) - Water passed from the atmosphere to the ground or plants above ground as rain or snow, or by condensation of water on plants.
prokaryote - A cellular organism lacking a membrane-bounded nucleus. Archaebacteria and eubacteria are prokaryotes.
protolith - An igneous or a sedimentary rock from which a metamorphic rock was transformed by metamorphism.
regolith - Disintegrated bedrock or gravel, sand, silt, clay, mud, and other materials at the surface of the earth that are not consolidated to form rock. Soils form within regolith.
rhizosphere - The area around roots that has special chemical properties and microbial

populations related to the roots.

saline - Saline soils and water are salty. They are alkaline if the dominant anion is carbonate or bicarbonate, but they may be neutral or acid if the dominant anions are chlorine or sulfate.

saprolite - weathered rock that has become soft without loss of bulk volume; it commonly retains some rock structure.

serpentinization - The hydrothermal alteration of Mg-silicate minerals to produce serpentine minerals, which are lizardite, chrysotile, and antigorite.

serpentinophile - A plant, or other kind of a living organism, that is more common in serpentine habitats than in other habitats.

siderophile (element) - a metallic element, a "lover" of iron (Greek *sideron*, iron, and *philia*, love) with less affinity than Cu and Zn for sulfur.

siderophore - A chemical compound with affinity for iron, specifically Fe(III).

silica - Silicon dioxide, SiO_2. It can be crystalline, as in quartz; cyrptocrystalline, as in chalcedony; or amorphous, as in opal.

silica boxwork - A network of veins where silica has been deposited in weathered ultramafic rock.

silica-carbonate rock - Rock consisting of variable quantities of silica (chalcedony and opal) together with magnesium carbonates and commonly stained rusty red by alteration of iron sulfide minerals. It forms erosion resistant rock outcrops with sparse vegetation.

soil, soils - The nonlithic outermost layer (meters thick, or less) of our planet, where reactions involve water and gases from the atmosphere and the actions of living organisms. Engineers commonly think of the entire regolith as soil.

soil classification - there are many systems of soil classification, but only one that is primarily for the USA. It is designated *Soil Taxonomy* (Soil Survey Staff 1999). It has four major levels, plus a family level and soil series within families.

specific gravity - The mass of a substance relative to the mass of water.

subduction - The descent of crust, generally oceanic crust, down into the mantle.

substrate (ecology) - The material inhabited by a living organism or on which an organism is growing.

talus - Coarse colluvium composed of rock fragments.

terrain (geoecology) - A physiographic landscape, or a landscape with limited ranges of rock stratigraphic, topographic, soil, and vegetation features.

terrane - A group of tectonically transported rocks or strata, as in accreted terranes, that share chronologic and geographic origins.

tetrahedron (mineralogy) - A three dimensional crystal structure with four triangular sides. The tetrahedra in phyllosilicates have oxygen ions at the apices, generally with a silicon ion in all tetrahedra of an tetrahedral sheet. Some small cations, such as aluminum, can replace silicon ions in tetrahedral positions.

transition elements - Chemical elements with unfilled d-shells, commonly including elements with filled d-shells, such as Cu and Zn, in which the valence electrons are in s-shells. They are the elements from Sc to Zn in the first transition, from Y to Cd in the second transition, and Hf to Hg in the third transition. They are transitional between elements of the first two columns on the left and elements of the last six columns on the right in the periodic table of elements (Appendix II).

transform fault - A strike-slip fault characteristic of oceanic ridges and along which spreading ridges are offset.

transpiration - The loss or water from photosynthesizing plants.

ultramafic rock - A rock composed chiefly of mafic or magnesium silicate minerals; for example, olivine, clinopyroxene, and orthopyroxene. Serpentinite is also an ultramafic rock.

vascular plant - A plant that has conductive tissues, such as xylem and phloem. Bryophytes are nonvascular plants.

vein (geology, petrography) - A thin fissure, or a crack filling. Chrysotile is common in veins within ultramafic rocks.

water, connate - water that was trapped in the pores of sedimentary rocks at the time that the sediments were deposited.

water, meteoric - Water that has circulated through the atmosphere.

weathering - the physical and chemical alteration of rock or regolith in contact with meteoric water or with air at atmospheric pressure.

References

Aalto, K.R. 2006. The Klamath Peneplain: A review of J.S. Diller's classic erosion surface. Geological Society of America, Special Paper 410: 451-463.

Aitken, J.D. 1959. Atlin map-area, British Columbia. Geological Survey of Canada, Memoir 307.

Alexander, E.B. 1995. Silica cementation in serpentine soils in the humid Klamath Mountains, California. Soil Survey Horizons 36: 154-159.

Alexander, E.B. 2003. Trinity Serpentine Soil Survey. Shasta-Trinity National Forest, Redding, California, unpublished manuscript, revised in 2004.

Alexander, E.B. 2004. Serpentine soil redness, differences among peridotite and serpentinite materials, Klamath Mountains, California. International Geology Review 46: 754-764.

Alexander, E.B. 2007. Baja California Soils with Ultramafic Parent Materials. Soil Survey Horizons 48(3): 67–70.

Alexander, E.B. 2009. Serpentine geoecology of the eastern and southeastern margins of North America. Northeastern Naturalist 16(5): 223-249.

Alexander, E.B. 2010. Old Neogene summer-dry soils with ultramafic parent materials. Geoderma 159: 2-8.

Alexander, E.B. 2011. Gabbro soils and plant distributions on them. Madroño 58: 113-122.

Alexander, E.B. 2013. Soils in Natural Landscapes. CRC Press, Boca Raton FL.

Alexander, E.B. 2014a. Arid to humid serpentine soils, mineralogy, and vegetation across the Klamath Mountains, USA. Catena 116: 114-122.

Alexander, E.B. 2014b. Foliar analyses of conifers on serpentine and gabbro soils in the Klamath Mountains. Madroño 61: 77-81.

Alexander, E.B. 2019. Ultramafic Aridisols on a sequence of fluvial terraces in Baja California. Geoderma Regional 17: article e00219.

Alexander, E.B., and J. Cooper DuShey. 2011. Topographic and soil differences from peridotite to serpentinite. Geomorphology 135: 271-276.

Alexander, E.B., P. Cullen, and P.J. Zinke. 1989. Soils and plant communities of ultramafic terrain on Golden Mountain, Cleveland Peninsula. Pages 47-56 *in* E.B. Alexander (editor). Proceedings of Watershed '89. USDA Forest Service, Alaska Region, Juneau, AK.

Alexander, E.B., C. Ping, and P. Krosse. 1994a. Podzolization in ultramafic materials in southeast Alaska. Soil Science 157: 46-52.

Alexander, E.B., R.C. Graham, C. Ping. 1994b. Cemented ultramafic till beneath a Podzol in Southeast Alaska. Soil Science 157: 53-58.

Alexander, E.B., R.G. Coleman, T. Keeler-Wolf, and S. Harrison. 2007a. Serpentine Geoecology of Western North America. Oxford University Press, New York.

Alexander, E.B., C.C. Ellis, and R. Burke. 2007b. A chronosequence of soils and vegetation on serpentine terraces in the Klamath Mountains, USA. Soil Science 172: 565-576.

Allen, B,L., and B,F. Hajek. 1989. Mineral occurrences in soil environments. Pages 199-278 *in* J.B. Dixon and S.B. Weed (editors). Minerals in Soil Environments. Soil Science Society of America, Madison, WI. p. 199-278.

Allison, J.E., G.W. Dittmar, and J.L. Hensell. 1975. Soil survey of Gillespie County, Texas. USDA, Natural Resources Conservation Service, Washington, DC.

Alvarado, G.E., P. Denyer, C.W. Sinton. 1997. The 89 Ma Tortugal komatiitic suite, Costa Rica: Implications for a common geological origin of the Caribbean and Eastern Pacific region from a mantle plume. Geology 25: 439-442.

Amato, J.M., J. Toro, E.L. Miller, G.E. Gehrels, G.L. Farmer, E.S. Gottlieb, and A.B. Till, 2009. Late Proterozoic-Paleozoic evolution of the Arctic Alaska-Chukotka terrane based on U-Pb igneous and detrital zircon ages: Implications for Neoproterozoic paleogeographic reconstructions: Geological Society of America, Bulletin 121: 1219-1235.

Arndt, N.T. 2008. Komatiite. Cambridge University Press, Cambridge, UK.

Arndt, N.T., and A.J. Naldrett. 1987. Komatiites in Munro Township, Ontario. Pages 317-322 *in* Geological Society of America, Centennial Field Guide, Northeastern Section.

Arroues, K. 2006. Soil Survey of Fresno County, California, Western Part. USDA, Natural Resources Conservation Service, Washington, DC.

Atzet, T., D.E. White, L.A. McCrimmon, P.A. Martinez, P.R. Fong, and V.D. Randal. 1996. Field guide to the forested plant associations of southwestern Oregon. USDA Forest Service R6-NR-ECOL-TP-17-96.

Bailey, E.H., W.P. Irwin, and D.L. Jones. 1964. Franciscan and related rocks, and their significance in the geology of western California. California Division of Mines and Geology, Bulletin 183.

Bailey, E.H., M.C. Blake, Jr., and D.L. Jones. 1970. On-land Mesozoic oceanic crust in California Coast Ranges. US Geological Survey, Professional Paper 700-C: 70-81.

Baker, A.J.M., and R.R. Brooks. 1989. Terrestrial higher plants that hyperaccumulate metal elements: a review of their distribution, ecology, and phytochemistry. Biorecovery 1: 81-126.

Baldwin, B.G., D.H. Goldman, D.J. Keil, R. Patterson, T.J. Rossatti, and D.H. Wilken (editors). 2012. The Jepson Manual, Vascular Plants of California. University of California Press, Berkeley.

Ballmer, M. 2008. Soil Survey of Santa Catalina Island, California. USDA, Natural Resources Conservation Service. Washington, DC.

Bangira, C. 2010. Mineralogy and Geochemistry of Soils of Ultramafic Origin from the Great Dyke, Zimbabwe and Gillespie County, Texas. PhD Dissertation, Texas A&M University.

Bangira, C., Y. Deng, R.H. Loeppert, E.T. Hallmark, and J.W. Stucki. 2011, Soil mineral composition in contrasting climatic regions of the Great Dyke, Zimbabwe. Soil Science Society of America, Journal 75: 2367-2378.

Barnes, I., V.C. LaMarche, Jr., and G. Himmelberg. 1967. Geochemical evidence of present-day serpentinization. Science 156: 630-632.

Barnes, I., J.R. O'Neil, and J.J. Trescases. 1978. Present day serpentinization in New Caledonia, Oman, and Yugoslavia. Geochemica et Cosmochimica Acta 42: 144-145.

Barros de Oliveira, S.M., J.J. Trescases, and A.J. Melfi. 1992. Lateritic nickel deposits of Brazil. Mineralium Deposita 27: 137-146.

Barton, A.M., and M.D. Wallenstein. 1997. Effects of invasion of *Pinus virginiana* on soil properties in serpentine barrens in southeastern Pennsylvania. Journal of the Torrey Botanical Society 124: 297-305.

Beck, R., and W.H. Weed. 1905. The Nature of Ore Deposits. The Engineering and Mining Journal. New York, London.

Begg, G.L., 1968. Soil Survey of Glenn County, California. US Government Printing Office,

Washington, DC.
Beinroth, F.H., 1982. Some highly weathered soils of Puerto Rico, 1. Morphology, formation, and classification. Geoderma 27: 1-72.
Bennett, H.H., and R.V. Allison. 1928. The soils of Cuba. Tropical Plant Research Foundation, Washington, DC.
Berazaín, R., 2001. The influence of serpentine soils on plants in Cuba. South African Journal of Science 97: 510-512.
Berazaín, R., 2004. Notes on tropical American nickel accumulating plants. Pages 255-258 *in* R.S. Boyd, A.J.M. Baker, and J. Proctor (editors), Ultramafic Rocks: their Soils, Vegetation, and Fauna. Science Reviews, St Albans, Herts, UK.
Best, M.G., 2003. Igneous and Metamorphic Petrology. Blackwell, Madden MA
Blake, M.C., and D.L. Jones. 1981. The Franciscan assemblage and related rocks in Northern California: a re-interpretation. Pages 306-328 *in* W.G. Ernst (editor). The Geotectonic Development of California. Prentice Hall, Englewood Cliffs, NJ.
Borchardt, G. 1989. Smectites. Pages 675-727 in J.B. Dixon and S.B. Weed (editors). Minerals in Soil Environments. Soil Science Society of America. Madison WI.
Borhidi, A. 1991. Phytogeography and Vegetation Ecology of Cuba. Akadémiai Kiadó, Budapest.
Borine, R. 1983. Soil Survey of Josephine County, Oregon. USDA, Soil Conservation Service.
Bostock, H.S. 1948. Physiography of the Canadian Cordillera, with special reference to the area north of the fifty-fifth parallel. Geological Survey of Canada, Memoir 247.
Bouchard, A,, H. Stuart, and E. Rouleau. 1978. Vascular flora of St. Barbe South district, Newfoundland: An interpretation based on biophysiographic areas. Rhodora 80: 228–308.
Boudette, E. 1982. Ophiolite assemblage of early Paleozoic age in central western Maine. Geological Association on Canada, Special Paper 24: 209–230.
Boyd, R.S. 2009. High-nickel insects and nickel hyperaccumulator plants; a review. Insect Science 16: 19-31.
Boyd, R.S., and S.N. Martens. 1994. Nickel hyperaccumulation by *Thlaspi montanum* var. *montanum* (Brassicaceae) is acutely toxic to insect herbivores. Oikos 70:21-25.
Boyd, R.S,, and S.N. Martens. 1998. The significance of metal hyperaccumulation for biotic interactions. Chemoecology 8: 1-7.
Boyd, R.S,, S.N. Martens, and M.A. Davis. 1999. The nickel hyperaccumulator *Streptanthus polygaloides* (Brassicaceae) is attacked by the parasitic plant *Cuscuta californica* (Cuscutaceae). Madroño 46: 92-99.
Boyd, R.S,, M.A. Wall, S.R. Santos, and M.A. Davis. 2009. Variations of morphology and elemental concentrations in the California nickel hyperaccumulator *Streptanthus polygaloides* (Brassicaceae). Northeastern Naturalist 16(5): 21-38.
Brand, N.W., C.R.M. Butt, and M. Elias. 1998. Nickel laterites: classification and features. AGSO Journal of Australian Geology and Geophysics 17: 81-88.
Brandon, M.T,, D.S. Cowan, and J.A. Vance. 1988. The Cretaceous San Juan thrust system, San Juan Islands, Washington. Geological Society of America, Special Paper 221: 1-83.
Briles, C.E., C. Whitlock, C.N. Skinner, and J. Mohr. 2011. Holocene forest development and maintenance on different substrates in the Klamath Mountains, northern California, USA. Ecology 92: 590-601.

Briscoe, L.R.E., T.B. Harris, W. Boussard, E. Dannenberg, F.C. Olday, and N. Rajakaruna. 2009. Bryophytes of adjacent serpentine and granite outcrops on the Deer Isles, Maine, USA. Rhodora 111: 1-20.

Brooks, R.R. 1987. Serpentine and Its Vegetation. Dioscorides Press, Portland, OR.

Brooks, R.R., J. Lee, R.D. Reeves, and T. Jaffé. 1977. Detection of nickeliferous rocks by analysis of herbarium specimens of indicator plants. Journal of Geochemical Exploration 7: 49-57.

Brown, E.H. 1977. Ophiolite on Fidalgo Island. Oregon Department of Geology and Mineral Industries, Bulletin 95: 67-74.

Brown, G.E., and T.P. Thayer. 1966. Geologic map of the Canyon City Quadrangle, northeastern Oregon. US Geological Survey, Miscellaneous Investigations, Map I-447.

Buchanan, F. 1807. A Journey from Madras through the Countries of Mysore, Canara, and Malabar. East India Company, London.

Bulmer, C.E., and L.M. Lavkulich. 1994. Pedogenic and geochemical processes of ultramafic soils along a climatic gradient in southwestern British Columbia. Canadian Journal of Soil Science 74: 165-177.

Bulmer, C.E., L.M. Lavkulich, and H.E. Schreier. 1992. Morphology, chemistry, and mineralogy of soils derived from serpentinite and tephra in southwestern British Columbia. Soil Science 154: 72-82.

Burch, S.H. 1968. Tectonic placement of the Burrow Mountain ultramafic body, Santa Lucia Range, California. Geological Society of America, Bulletin 60: 527-544.

Burgess, J., K. Szlavecz, N. Rajakaruna, S. Lev, and C. Swan. 2015. Vegetation dynamics and mesophication in response to conifer encroachment within an ultramafic system. Australian Journal of Botany, dx.doi.org/10.1071/BT14241.

Burns, L.E. 1985. The Border Ranges ultramafic and mafic complex, south-central Alaska: cumulate fractionates of island-arc volcanics. Canadian Journal of Earth Sciences 22: 1020-1038.

Burt, R. (Editor). 2004. Soil Survey Laboratory Methods Manual. USDA, Natural Resources Conservation Service, Soil Survey Investigations Report No. 42.

Burt, R., M. Fillmore, M.A. Wilson, E.R. Gross, R.W. Langridge, and D.A. Lammers. 2001. Soil properties of selected pedons on ultramafic rocks in Klamath Mountains, Oregon. Communications in Soil Science and Plant Analysis 32: 2145-2175.

Butler, J.R. 1989. Review and classification of ultramafic bodies in the Piedmont of the Carolinas. Geological Society of America, Special Paper 231: 19–31.

Butt, C.A.M., and D. Cluzel. 2013. Nickel laterite ore deposits: weathered serpentinite. Elements 8(2); 123-238.

Caillaud, J., D. Proust, and D. Righi, 2006. Weathering sequences of rock-forming minerals in a serpentinite: influence of microsystems on clay mineralogy. Clays and Clay Minerals 54: 87-100.

Caillaud, J,, D. Proust, S, Philippe, C, Fontaine, and M, Fialin. 2009. Trace metals distribution from a serpentine weathering at the scales of the weathering profile and its related weathering microsystems and clay minerals. Geoderma 149: 199-208.

Campbell, A.R., C. Stone, N. Shamsedin, D. Kolterman, and A. Pollard. 2013. Faculative hyperaccumulation of nickel in *Psychotria grandis* (Rubiaceae). Caribbean Naturalist 1: 1-8.

Camprubi, A. 2017. The metallogenic evolution in Mexico during the Mesozoic, and its bearing in the Cordillera of western North America. Ore Geology Reviews 81: 1193-1214.
Camuti, K.S., and M.G. Gifford. 1997. Mineralogy of the Murrin Murrin nickel laterite deposit, Western Australia. Pages 407-410 *in* H. Papunen (editor). Mineral deposits: Research and Exploration. Balkema, Rotterdam.
Cannings, S., J.A. Nelson, and R. Cannings. 2011. Geology of British Columbia, A journey through time. Greystone Books, Vancouver, BC.
Carlson, R.W. 2019. Analysis of lunar samples: Implications for planet formation and evolution. Science 365: 340-344.
Cater, F.W., and F.G. Wells. 1953. Geology and mineral resources of the Gasquet Quadrangle, California-Oregon. US Geological Survey, Bulletin 558-C.
Cedeño-Maldonado, J.A., and G.J. Breckon. 1996. Serpentine endemism in the flora of Puerto Rico. Caribbean Journal of Science 32: 348.
Centeno-Garcia, E. 2017. Mesozoic tectono-magmatic evolution of Mexico: an overview. Ore Geology Reviews 81: 1035-1052.
Centeno-Garcia, E., C. Busby, M. Busby, and G. Gehrels. 2011. Evolution of the Guerrero composite terrane along the Mexican margin, from extensional fringing arc to contractional continental arc. Geological Society of America, Bulletin 123: 1776-1797.
Cerpa, A., M.P. Garcia-Gonzales, P. Taraj, J. Requena, L. Garcell, and C.J. Serna. 1999. Mineral content and particle-size effects on the colloidal properties of concentrated lateritic suspensions. Clays and Clay Minerals 57: 515-521.
Chang, S. 2004. Forest Service Research Natural Areas in California. US Department of Agriculture, General Technical Report PSW-GTR-188.
Chen, M., and L.Q. Ma, 2001. Comparison of three aqua regia digestion methods for twenty Florida soils. Soil Science Society of America, Journal 65, 491-499.
Churchill, D. 1988. Soil Survey of Plumas National Forest Area, California. USDA, Forest Service, Pacific Southwest Region, Vallejo, California.
Churchill, R.K., and R. Hill. 2000. A general location guide for ultramafic rocks in California - Areas more likely to contain naturally occurring asbestos. California Division of Mines and Geology. Open File Report 2000-19.
Cleaves, E.T., D.W. Fisher, and O.P. Bricker. 1974. Chemical weathering of serpentinite in the eastern Piedmont of Maryland. Geological Society of America, Bulletin 85: 437-444.
Coleman, R.G. 1996. New Idria serpentinite: A management dilemma. Environmental and Engineering Geoscience 2: 9-22.
Coleman, R.G. 2004. Geologic nature of the Jasper Ridge Biological Preserve, San Francisco peninsula, California. International Geology Review 46: 629-637.
Colpron, M., J.L. Nelson, and DC. Murphy. 2007. Northern Cordilleran terranes and their interactions through time. GSA Today 17(4/5): 4-10.
Colwell, W., W. McGee, and G. McClellan. 1955. Vegetation-Soil Maps, quadrangles 59B-1,2,3,4, Lake County. California Department of Forestry, Sacramento.
Condie, K.C. 2005. Earth as an Evolving Planetary System. Elsevier, Amsterdam.
Cook, T.D. 1978. Soil Survey of Monterey County, California. US Government Printing Office, Washington, DC.
Dahlgren, R.A. 1994. Soil acidification and nitrogen saturation from weathering of ammonium-

bearing rock. Nature 368: 838-841.

D'Amico, M.E., F. Julitta, D. Cantelli, and F. Previtali. 2008. Podzolization over ophiolitic materials in the western Alps (Natural Park of Mont Avic, Aosta Valley, Italy). Geoderma 146:129–136.

Damschen, E.I., S. Harrison, and J.B. Grace. 2010. Climate change effects on an endemic rich flora: resurveying Robert H. Whittaker's sites (Oregon, USA). Ecology 91: 3609-3619.

Dauphin, G, and M.H. Grayum. 2005. Bryophytes of the Santa Elena Peninsula and Islas Murciélago, Guanacaste, Costa Rica, with special attention to neotropical dry forest habitats. Lankesteriana 5(1): 53-61.

Day, H.W., E.M. Moores, and A.C. Tuminas. 1985. Structure and tectonics of the northern Sierra Nevada. Geological Society of America, Bulletin 96: 436-450.

Dayton, B.R. 1966. The relationship of vegetation to Iredell and other Piedmont soils in Granville County, North Carolina. Journal of the Elisha Mitchell Scientific Society 82: 108–118.

Dearden, P. 1979. Some factors influencing the composition and location of plant communities on a serpentine bedrock in western Newfoundland. Journal of Biogeography 6: 93–104.

Deer, W.A., R.A. Howie, and J. Zussman. 1966. An Introduction to the Rock Forming Minerals. Wiley, New York.

De Kimpe, C., and J. Zizka. 1973. Weathering and clay formation in a dunite deposit at Asbestos. Canadian Journal of Soil Science 10: 27–35.

De Kimpe, C., J. Dejou, and Y. Chevalier. 1987. Évolution géochimique superficielle des pyroxénites ignées du Mont Saint-Bruno, Québec. Canadian Journal of Soil Science 24: 760–770.

DeLapp, J.A., and B.F. Smith. 1978. Soil-Vegetation Maps, quadrangles 61D-1, 2, 3, 4), Sonoma County. California Department of Forestry, Sacramento.

Delvigne, J., E.B.A. Bisdom, J. Sleeman, and G. Stoops. 1979. Olivines, their pseudomorphs and secondary products. Pedologie 29(3): 247-309.

Denny, C.S. 1982. Geomorphology of New England. US Geological Survey, Professional Paper 1208.

Dickinson, W.R. 1966. Table Mountain serpentinite extrusion in the California Coast Ranges. Geological Society of America, Bulletin 77: 451-472.

Dickinson, W.R., and T.A.L. Casey. 1976. Sedimentary serpentine of the Miocene Big Blue formation near Cantua Creek, California. Pages 65-74 *in* A.E. Frittsche, H. Ter Best, Jr., and N.W. Wornardt. Tectonic and Geological History of the Pacific Coast of North America. Society of Economic Paleontologists and Mineralogists, Pacific Section.

Dickinson, W.R, and T.F. Lawton. 2001. Carboniferous to Cretaceous assembly and fragmentation of Mexico. Geological Society of America, Bulletin 113: 1142-1160.

Dickinson, W.R., C.A. Hopson, and J.B. Saleeby. 1996. Alternate origins of the Coast Range ophiolite (California): Introduction and implications. GSA Today 6(2): 1-10.

Diller, J.S. 1902. Physiographic development of the Klamath Mountains. US Geological Survey, Bulletin 196.

Dirven, J.M.C., J. Van Schuylenborgh, and N. van Breeman. 1976. Weathering of serpentine in Matanzas Province, Cuba: mass transfer calculations and irreversible reaction pathways. Soil Science Society of America, Journal 40: 901-907.

Dittemore, W.H. Jr., and J.E. Allison. 1979. Soil survey of Blanco and Burnet counties, Texas. USDA, Natural Resources Conservation Service, Washington, DC.

Dorsey, R.J., and T.A. LaMaskin. 2007. Stratigraphic record of Triassic-Jurassic collision tectonics in the Blue Mountains province, northeastern Oregon. American Journal of Science 307: 1167-1193.

Drees, L.R., L.P. Wilding, N.E. Smeck, and A.L. Senkayi. 1989. Silica in soils: quartz and disordered silica polymorphs. Pages 913-974 *in* J.B. Dixon and S.B. Weed (editors). Minerals in Soil Environments. Soil Science Society of America, Madison WI.

Dukes, J.S. 2001. Productivity and complementarity in grassland microcosms of varying diversity. Oikos 94: 468-480.

Dunning, D. 1942. A site classification for the mixed conifer selection forests of the Sierra Nevada. Research Note No. 28. US Department of Agriculture, Forest Service, California Forest and Range Experiment Station, Berkeley CA.

Edwards, S.W. 1994. Bear Valley: wildflowers as John Muir described them. Fremontia 22(4): 12-16.

Ehrlich, P.R., and I. Hanski. 2004. On the Wings of the Checkerspots: A Model System for Population Biology. Oxford University Press, New York.

Ertter, B., and M.L. Bowerman. 2002. The Flowering Plants of Mount Diablo. California Native Plant Society, Sacramento, CA.

Estrada, S,, K. Mende, A. Gerdes, A. Gärtner, M. Hofmann, C. Spiegel, D. Damaske, and N. Koglin. 2018. Proterozoic to Cretaceous evolution of the western and central Pearya terrane (Canadian High Arctic). Journal of Geodynamics 120: 45-76.

Eswaran, H., A. van Wambeke, and F.H. Beinroth. 1979. A study of some highly weathered soils of Puerto Rico, morphological properties. Pedologie 29(2): 139-162.

Evens, J., and S. San. 2004. Vegetation associations of a serpentine area: Coyote Ridge, Santa Clara County, California. Unpublished report, California Native Plant Society, Sacramento.

Eyles, N., and A. Miall. 2007. Canada Rocks. Fitzhenry & Whiteside, Brighton, MA.

FAO/ISRIC/ISSS. 1998. World Reference Base for Soil Resources. World Soil Resources Report No. 84. Food and Agricultural Organization of the United Nations, Rome.

Ferrians J.C. 1965. Permafrost Map of Alaska. US Geological Survey, Miscellaneous Geological Investigations Map I-445.

Fiedler, P.L., and R.A. Leidy. 1987. Plant communities of the Ring Mountain Preserve, Marin County, California. Madroño 34: 173-196.

Fillmore, M.H. 2005. Soil Survey of Curry County, Oregon. USDA, Natural Resources Conservation Service, Washington DC.

Findlay, D.C. 1969. Origin of the Tulameen ultramafic gabbro complex, southern British Columbia. Canadian Journal of Earth Science 6: 399-425.

Foose, M.P. 1992. Nickel–mineralogy and chemical composition of some nickel-bearing laterites in southern Oregon and northern California. US Geological Survey, Bulletin 1877E.

Fox, RL. 1982. Some highly weathered soils of Puerto Rico, 3. Chemical properties. Geoderma 27: 139-176.

Garrison, J.R, Jr. 1981. Coal Creek serpentinite, Llano uplift, Texas: A fragment of an incomplete Precambrian ophiolite. Geology 9: 225–230.

Goldin, A. 1992. Soil Survey of Whatcom County Area, Washington. US Natural Resource

Conservation Service. Washington, DC.
Gómez-Gómez, F,, J. Rodrigues-Martinez, and M. Santiago. 2014. Hydrogeology of Puerto Rico, and the Outlying Islands of Vieques, Culebra, and Mona. US Geological Survey, Scientific Investigations Map 3296.
Gowans, K.D. 1967. Soil Survey of Tehama County, California. US Government Printing Office, Washington, DC, 124 pages, maps.
Graham, R.C., M.M. Diallo, and L.J. Lund. 1990. Soils and mineral weathering on phyllite colluvium and serpentine in northwestern California. Soil Science Society of America, Journal 54: 1682-1690.
Graham, R.C., and A.T. O'Geen. 2010. Soil mineralogy trends in California landscapes. Geoderma 154: 418-437.
Gram, W.K. , E. Borer, K. Cottingham, E. Seabloom, V. Boucher, L. Goldwasser, F. Micheli, B. Kendall, and R. Burton. 2004. Distribution of plants in a California serpentine grassland: are rocky hummocks spatial refuges for native species? Plant Ecology 172: 159-171.
Gray, J.T., and R.J.E. Brown. 1979. Permafrost existence and distribution in the Choc-Chocs Mountains, Gaspésie, Québec. Géographie Physique et Quaternaire 33: 299–316.
Griffin, J.R.. 1975. Plants of the highest Santa Lucia and Diablo Range peaks, California. USDA, Forest Service, Research Paper 110.
Hack, J.T. 1982. Physiographic divisions and differential uplift in the Piedmont and Blue Ridge. US Geological Survey, Professional Paper 1265.
Hanan, B.B., and A.K. Sinha. 1989. Petrology and tectonic affinity of the Baltimore mafic complex, Maryland. Geological Society of America, Special Paper 231: 1–18.
Hanes, R.O. 1994. Soil Survey of Tahoe National Forest Area, California. USDA, Forest Service, Pacific Southwest Region, Vallejo, California.
Harris, R.A. 1998. Origin and tectonic evolution of the metamorphic sole beneath the Brooks Range ophiolite. Geological Society of America, Special Paper 324: 293-312.
Harris, T.B., F.C. Olday, and N. Rajakaruna. 2007. Lichens of Pine Hill, a peridotite outcrop in eastern North America. Rhodora 109: 430–447.
Harris, W.G., L.W. Zelazny, and J.C. Baker. 1984. Depth and particle size distribution of talc in a Virginia piedmont Ultisol. Clays and Clay Minerals 32: 227–230.
Harrison, S. 1997. How natural habitat patchiness affects the distribution of diversity in Californian serpentine chaparral. Ecology 78: 1898-1906.
Harrison, S. 1999. Local and regional diversity in a patchy landscape: native, alien and endemic herbs on serpentine soils. Ecology 80: 70-80.
Harrison, S. 2013. Plant and Animal Endemism in California. University of California Press, Berkeley.
Harrison, S., and N. Rajakaruna, (editors). 2011. Serpentine: the Evolution of a Model System. University of California Press, Berkeley.
Harrison, S., D.D. Murphy, and P.R. Ehrlich. 1988. Distribution of the Bay checkerspot butterfly, *Euphydryas editha bayensis*: evidence for a metapopulation model. American Naturalist 132: 360-382.
Harrison, S., J.L.Viers, and J.F. Quinn. 2000. Climatic and spatial patterns of diversity in the serpentine plants of California. Diversity and Distributions 6: 153-161.
Harrison, S., H.D. Safford, J.B. Grace, J.H. Viers, and K.F. Davies. 2006. Regional and local

species richness in an insular environment: serpentine plants in California. Ecological Monographs 76: 41-56.

Hastings, J.R., and R.R. Humphrey. 1969. Climatological Data and Statistics for Baja California. University of Arizona, Institute of Atmospheric Physics, Technical Reports for Meteorology and Climatology, No. 18, 75 pages.

Hastings, J.R., and R.M. Turner. 1965. Seasonal Precipitation Regimes in Baja California. Geografiska Annaler 47A: 204-223.

Hatch, N.L., Jr. 1982. Taconian line in western New England and its implications to Paleozoic tectonic history. Geological Association on Canada, Special Paper 24: 67–85.

Heald, W.R. 1965. Calcium and magnesium. Pages 999-1010 *in* C.A. Black (editor), Methods of Soil Analysis, Part 2, Chemical and Microbiological Properties. American Society of Agronomy, Madison WI.

Hendriksen, N. 2008. Geological History of Greenland – Four Billion Years of Earth Evolution. Geological Survey of Denmark and Greenland, Copenhagen.

Hershey, O.H. 1903. The relation between certain river terraces and the glacial series in northwestern California. Journal of Geology 11: 431-458.

Hibbard, J.P., C.R. van Staal, D.W. Rankin, and H. Williams. 2006. Lithotectonic map of the Appalachian orogen, Canada-United States of America. Geological Survey of Canada, Map 2096 A.

Hickman, J.C. (editor). 1993. The Jepson Manual: higher plants of California. University of California Press, Berkeley.

Hietanen, A. 1973. Geology of the Pulga and Bucks Lake Quadrangles, Butte and Plumas Counties, California. US Geological Survey, Professional Paper 731.

Higgins, A.K., J.A. Gilotti, and M.P. Smith (editors). 2008. The Greenland Caledonides. Geological Society of America, Memoir 202.

Himmelberg, G.R., and R.A. Loney. 1980. Petrology of ultramafic and gabbroic rocks of the Canyon Mountain ophiolite, Oregon. American Journal of Science 280-A: 232-268.

Himmelberg, G.R., and R.A. Loney. 1995. Characteristics and petrogenesis of Alaskan-type ultramafic-mafic intrusions, southeastern Alaska. USGS Professional Paper 1564: 1-47.

Hitchcock, C.L., and A. Cronquist. 1973. Flora of the Pacific Northwest. University of Washington Press, Seattle.

Hobbs, R.J., and H.A. Mooney. 1995. Spatial and temporal variability in California annual grassland: results from a long-term study. Journal of Vegetation Science 6: 43-56.

Hochman, D.J. 2001. *Pinus virginiana* invasion and soil-plant relationships of Soldier's Delight Natural Environmental Area, a serpentine site in Maryland. M.C. Thesis. University of Maryland, College Park, MD.

Holland, S.S. 1976. Landforms of British Columbia, A Physiographic Outline. British Columbia Department of Mines and Petroleum Resources, Bulletin 48.

Holmgen, G. 1967. A rapid citrate-dithionite extractable iron procedure. Soil Science Society of America, Proceedings 31: 210-211.

Hopson, C.A., J.M. Mattson, E.A. Pessagno, and B.P. Luyendyk. 2008. California Coast Range ophiolite: Composite Middle and Early Jurassic oceanic lithosphere. Geological Society of America, Special Paper 438: 1-101.

Hotz, P.E. 1964. Nickeliferous laterites in southwestern Oregon and northwestern California.

Economic Geology 59: 355-396.
Hubers, H., S, Borges, and M, Alfaro. 2003. The oligochaetofauna of the Nipe soils in the Maricao State Forest. Pedobiologia 47: 475-478.
Huggett, R,J. 1995. Geoecology. Routledge, London.
Hull, J.C., and S,G. Wood. 1984. Water relations of oak species on and adjacent to a Maryland serpentine soil. American Midland Naturalist 112: 224-234.
Hultén, E. 1968. Flora of Alaska and neighboring territories: a manual of the vascular plants. Stanford University Press, Stanford, CA.
Hunter, J.C., and J.E. Horenstein. 1992. The vegetation of the Pine Hill area (California) and its relation to substratum. Pages 197-206 *in* A.J.M. Baker, J. Proctor, and R. Reeves (editors). The Vegetation of Ultramafic (Serpentine) Soils. Intercept, Andover, Hampshire, UK.
Irvine, T.N. 1974. Petrology of the Duke Island ultramafic complex, southeastern Alaska. Geological Society of America, Memoir 138.
Irvine, T.N., and C.H. Smith. 1967. The ultramafic rocks of the Muskox intrusion, Northwest Territories, Canada. Pages 38-49 *in* P.J. Wyllie (editor). Ultramafic and Related Rocks. Wiley, New York.
Irwin, W.P. 1960. Geologic reconnaissance of the northern Coast Ranges and Klamath Mountains, California. California Division of Mines, Bulletin 179: 1-80, map.
Irwin, W.P. 1977. Ophiolitic terranes of California, Oregon, and Nevada. Pages 75-92 *in* R.G. Coleman and W.P. Irwin (editors). North American Ophiolites. Oregon Department of Geology and Mineral Industries, Bulletin 95.
Irwin, W.P. 1981. Tectonic accretion of the Klamath Mountains. Pages 29-49 *in* W.G. Ernst (editor), The Geotectonic Development of California. Prentice-Hall, Englewood Cliffs, NJ.
Istok, J.D., and M.E. Harward. 1982. Influence of soil moisture on smectite formation in soils derived from serpentine. Soil Science Society of America, Journal 46: 1106-1108.
Jaffré, T., R.R. Brooks, J. Lee, and R.D. Reeves. 1976. *Sebertia acuminata*: a hyper-accumulator of nickel from New Caledonia. Science 193: 579-580.
James, O.B. 1971. Origin and emplacement of the ultramafic rocks of the Emigrant Gap area, California. Journal of Petrology 12: 523-560.
Jennings, C.W. 1977. Geologic Map of California. The Resources Agency, Department of Conservation, Division of Mines and Geology, State of California.
Jenny, H. 1980. The Soil Resource. Springer-Verlag, Berlin.
Jensen, S., and D.R. Pyke. 1982. Komatiites in the Ontario portion of the Abitibi belt. Pages 147-157 *in* N.T. Arndt and E.G. Nesbit (editors). Komatiites.
Jhee, E.M., R.S. Boyd, and M.D. Eubanks. 2005. Nickel hyperaccumulation as an elemental defense of *Streptanthus polygaloides* (Brassicaceae): influence of herbivore feeding mode. New Phytologist 168: 331-344.
Jimerson, T.M., L.D. Hoover, E. McGee, G. DeNitto, and R.M. Creasy. 1995. A Field Guide to Serpentine Plant Associations and Sensitive Plants in Northwestern California. USDA Forest Service R5-ECOL-TP-006.
Johnson, D.R., J.T. Haagen, and A.C. Terrell. 2003. Soil Survey of Douglas County Area, Oregon. USDA, Natural Resources Conservation Service, Washington DC..
Jones, R.C., W.H. Hudnall, and W.S. Sakai. 1982. Some highly weathered soils of Puerto Rico, 2 Mineralogy. Geoderma 27: 75-137.

Jones, W.E. 1962. Soil Survey of Cherokee County, South Carolina. US Department of Agriculture, Washington DC.
Juday, G.P, 1992. Alaska Research Natural Areas. 3: Serpentine Slide. General Technical Report PNW-GTR-271, USDA, Forest Service.
Jurjavcic, N., S. Harrison, and A. Wolf. 2002. Abiotic stress, competition, and the distribution of the native grass *Vulpia microstachys* in a mosaic environment. Oecologia 130: 555-562.
Kaplan, T. 1984. The Lassics outlier, an outlier of Coast Range ophiolite, northern California. SEPM, Pacific Division v. 43 (Franciscan Geology of Northern California): 203-219.
Kashiwagi, J.H. 1985. Soil Survey of Marin County. US Government Printing Office, Washington, DC.
Kazakou, E., P.G. Dimitrakopoulos, A.J.M. Baker, R.D. Reeves, and A.Y. Troumbis. 2008. Hypotheses, mechanisms and tradeoffs of tolerance and adaptation to serpentine soils: From species to ecosystem level. Biological Reviews 83: 495-508.
Keeler-Wolf, T. 1983. An Ecological Survey of the Frenzel Creek Research Natural Area, Mendocino National Forest, Colusa County, California. USDA Forest Service, Pacific Southwest Region, Vallejo, CA, unpublished report.
Keeler-Wolf, T. 1990. Ecological surveys of Forest Service Research Natural Areas in California. US Department of Agriculture, General Technical Report PSW-GTR-125.
Keppie, J.D. 2004. Terranes of Mexico revisited: a 1.3 billion year odyssey. International Geology Review 26: 765-794.
Kerans, C. 1983. Timing of emplacement of the Muskox intrusion: constraints from Coppermine homocline cover strata. Canadian Journal of Earth Sciences 20: 673-683.
Kerr, A.C., G.F. Marriner, N.T. Arndt, J. Tarney, A. Nivia, A.D. Saunders, and R.A. Duncan. 1996. The petrogenesis of Gorgona komatiites, picrites, and basalts: new field, petrographic, and geochemical constraints. Lithos 37: 245-260.
Klungland, M.W., and M. McArthur. 1989. Soil Survey of Skagit County Area, Washington. USDA, Soil Conservation Service. Washington, DC
Kram, P., J. Hruška, and J.B. Shanley. 2012. Stream geochemistry in three contrasting monolithic Czech watersheds. Applied Geochemistry 27: 1854-1863.
Krasilnikov, P., Ma del C. Gutiérrez-Castorena, R.J. Ahrens, C.O. Cruz-Gaistado, S. Sedov, and E. Solleirra-Robelledo. 2013. The Soils of Mexico. Springer.
Kruckeberg, A.R. 1951. Intraspecific variability in the response of certain native plant species to serpentine soil. American Journal of Botany 33: 408-419.
Kruckeberg, A.R. 1957. Variation in fertility of hybrids between isolated populations of the serpentine species, *Streptanthus glandulosa* Hook. Evolution 11: 185-211.
Kruckeberg, A.R. 1967. Ecotypic response to ultramafic soils by some plant species of northwestern North America. Brittonia 19:133-151.
Kruckeberg, A.R. 1969. Plant life on serpentinite and other ferromagnesian rocks in northwestern North America. Syesis 2: 15-114.
Kruckeberg, A.R. 1979. Plants that grow on serpentine - A hard life. Davisonia 10: 21-29.
Kruckeberg, A.R. 1984. California Serpentines: Flora, Vegetation, Geology, Soils, and Management Problems. University of California Press, Berkeley.
Kruckeberg, A.R., and R.D Reeves. 1995. Nickel hyperaccumulation by serpentine species of *Streptanthus* (Brassicaceae): Field and greenhouse studies. Madroño 42: 458-469.

Lambert, G., J. Kashiwagi, B. Hansen, P. Gale, and A. Endo. 1978. Soil Survey of Napa County. US Government Printing Office, Washington, DC.
Langmuir, D. 1965. Stability of carbonates in the system MgO-CO_2-H_2O. Journal of Geology 73: 730-754.
Lanspa, K.E. 1993. Soil Survey of Shasta-Trinity Forest Area, California. USDA, Forest Service, Shasta-Trinity National Forest, Redding, CA,
Laó-Dávila, D.A., P.A. Lierandi-Román, and T.H. Anderson. 2012. Cretaceous-Paleogene thrust emplacement of serpentinite in southwestern Puerto Rico. Geological Society of America, Bulletin 124: 1169-1190.
Larabee, D.M. 1966. Map showing distribution of ultramafic and intrusive mafic rocks from northern New Jersey to Alabama. US Geological Survey, Map I–476.
Larabee, D.M. 1971. Map showing distribution of ultramafic and intrusive mafic rocks from New York to Maine. US Geological Survey, Map I–676.
Lazarus, B.E., J.H. Richards, V.P. Claassen, R.E. O'Dell, and M.A. Ferrell. 2011. Species specific plant-soil interactions influence plant distribution on serpentine soils. Plant and Soil 342: 327-344.
Lee, B.D., S.K. Sears, R.C. Graham, C. Amrhein, and H. Vali. 2003. Secondary mineral genesis from chlorite and serpentine in an ultramafic soil toposequence. Soil Science Society of America, Journal 67: 1309-1317.
Le Maitre, R.W. 1976. The chemical variability of some common igneous rocks. Journal of Petrology 17: 589-637.
Le Maitre, R.W. (editor). 2002. Igneous Rocks: A Classification and Glossary of Terms. Cambridge University Press, Cambridge, UK.
Lessovaia S., S. Dultz, Y. Polekhovsky, V. Krupskaya , M. Vigasina , and L. Melchakova. 2012. Rock control of pedogenic clay mineral formation in a shallow soil from serpentinous dunite in the Polar Urals, Russia. Applied Clay Science 64: 4-11.
Lewis, G.J., and G.E. Bradfield. 2003. A floristic and ecological analysis at the Tulameen ultramafic (serpentine) complex, southern British Columbia, Canada. Davidsonia 14: 121-128, 131-134, 137-144.
Lewis, G.J., and G.E. Bradfield. 2004. Plant community-soil relationships at an ultramafic site in southern British Columbia, Canada. Pages 191-197 *in* R.S. Boyd, A.J.M. Baker, and J. Proctor (editors). Ultramafic Rocks: Their Soils, Vegetation, and Fauna. Science Reviews, St. Albans, Herts, UK.
Lewis, G.J., J.M. Ingram, and G.E. Bradfield. 2004. Diversity and habitat relationships of bryophytes at an ultramafic site in southern British Columbia, Canada. Pages 199-204 *in* R.S. Boyd, A.J.M. Baker, and J. Proctor (editors). Ultramafic Rocks: Their Soils, Vegetation, and Fauna. Science Reviews, St. Albans, Herts, UK.
Lewis, J.F., G. Draper, J.A. Proenza, J. Espaillat, J. Jiménez. 2006. Ophiolite-related rock (serpentinite) in the Caribbean region: a review of the occurrence, composition, origin, emplacement, and Ni-laterite soil formation. Geologica Acta 4: 237-263.
Lohnes, R.A., and T. Demirel. 1973. Strength and structure of laterites and lateritic soils. Engineering Geology 7: 13-33.
Loney, R.A., and G.R. Himmelberg. 1988. Ultramafic rocks of the Livengood terrane. US Geological Survey, Circular 1016: 68-70.

Lugo-Camacho, J.L.. 2008. Soil Survey of San Germán Area, Puerto Rico. USDA, Natural Resources Conservation Service.
Lugo-López, M.A., J.M. Wolf, and R. Perez-Escolar. 1981. Water-loss, intake, movement, retention, and availability in major soils of Puerto Rico. Puerto Rico Agriculture Experiment Station, Bulletin 264: 1-24.
Lyons, J.B., E.L. Boudette, and J.N. Aleinikoff. 1982. The Avalon and Gander Zones in central eastern New England. Geological Association of Canada, Special Paper 24:43–65.
MacDonald, J.S. Jr., G. Harper, R. Miller, J. Miller, A. Mlinarevich, and and C. Schultz. 2008. The Ingalls ophiolite complex, central Cascades, Washington: geochemistry, tectonic setting, and regional correlations. Geological Society of America, Special Paper 438: 133-159.
Malcolm, B., N. Malcolm, J. Shevock, and D. Norris. 2009. California Mosses. Micro-Optics Press, Nelson, NZ.
Mankinen, E.A., and W.P. Irwin. 1990. Review of data from the Klamath Mountains, Blue Mountains, and Sierra Nevada: Implications for paleogoegraphic reconstructions. Geological Society of America, Special Paper 255: 397-409.
Mann, P. 2007. Overview of the tectonic history of northern Central America. Pages 1-19 *in* Geological and Tectonic Development of the Caribbean Plate Boundary in Northern Central America. Geological Society of America, Special Paper 428.
Mansberg, L., and T.R. Wentworth. 1984. Vegetation and soils of a serpentine barren in western North Carolina. Torrey Botanical Club, Bulletin 111: 273–286.
Maoui, H.M. 1966. A Mineralogical and Genetic Study of Serpentine Derived Soils in Gillespie County, Texas. M.S. Thesis, Texas Tech University, Lubbock, TX.
Marrero-Rodríguez, A., J.M. Pérez-Jeménez, E. Suárez-Estrada, and E. Vegas-Lorenzo. 1989. Paginas IX.1-IX.3, Mapa de Suelos 1:1 000 000, *in* Atlas Nacional de Cuba, Academia de Ciencias de Cuba.
Marschner, H. 2002. Mineral Nutrition of Higher Plants. Academic Press, London.
Masan, B.O., and J. Kushiro. 1977. Compositional variations of coexisting phases with degree of melting in peridotite in the upper mantle. American Mineralogist 62: 843-865.
Mathewes, R.W., and J.A. Westgate, 1980. Bridge River tephra: revised distributions and significance for detecting old carbon errors in radiocarbon dates of limnic sediments in southern British Columbia. Canadian Journal of Earth Sciences 17: 1454-1461.
Mathews, Bill, and Jim Monger. 2005. Roadside Geology of Southern British Columbia. Mountain Press, Missoula MT.
Matsusaka, Y., G. Sherman, and D. Swindale. 1965. Nature of magnetic minerals in Hawaiian soils. Soil Science 100: 192-199.
McCallum, L.S. 1996. The Stillwater complex. Pages 441-483 *in* P.G. Cawthorn (editor). Layered Intrusions. Elsevier, Amsterdam.
McCarten, N. 1992. Community structure and habitat relations in a serpentine grassland in California. Pages 207-211 *in* A.J.M. Baker, J. Proctor, and R.D. Reeves (editors). The Vegetation of Ultramafic (Serpentine) Soils. Intercept, Andover, NH, UK.
McGahan, D.G., R.J. Southard, and V.P. Claassen. 2008. Tectonic inclusions in serpentinitic landscapes contribute to plant nutrient calcium. Soil Science Society of America, Journal 72:838-847.
McLaughlin, R., M. Blake, A. Griscom, C. Blome, B. Murchey. 1988. Tectonics of formation,

translation, and dispersal of the Coast Range ophiolite of California,.Tectonics 7: 1033-1039.

McLelland, J.M., B.W. Selleck, and M.E. Bicford. 2010. Review of the Proterozoic evolution of the Grenville Province, its Adirondack outlier, and the Mesoproterozoic inliers of the Appalachians. Pages 1-29 *in* R.P. Tollo, M.J. Bartholomew, J.P. Hibbard, and P.M. Karabinos (editors). From Rodinia to Pangea: the Lithotectomic Record of the Appalachian Region. Geological Society of America, Memoir 206.

Medeiros, J.D., A.M. Fryday, and N. Rajakaruna. 2014. Additional lichen records and mineralogical data from metal-contaminated sites in Maine. Rhodora 116: 323-347.

Medeiros, J.D., N. Rajakaruna, and E.B. Alexander. 2015. Gabbro soil-plant relations in the California Floristic Province. Madroño 2: 75-87.

Medina, E., E. Calves, J. Figaro, and E. Lugo. 1994. Mineral content of leaves from trees growing on serpentine soils under contrasting rainfall regimes in Puerto Rico. Plant and Soil 158: 13-21.

Merrill, G.P. 1906. A Treatise on Rocks, Rock-weathering, and Soils. Macmillan, New York.

Metcalfe, R.V., and J.W. Shervais. 2008. Suprasubduction-zone ophiolites: Is there really an ophiolite conundrum? Geological Society of America, Special Paper 138: 191-222.

Miller, V.C. 1972. Soil Survey of Sonoma County, California. US Government Printing Office, Washington, DC.

Miller, W.R., D.B. Smith, E. Meier, P.H. Briggs, P.M. Theodorakos, R.F. Sanzolone, and R.B. Vaughn. 1998. Geochemical Baselines for Surface Waters and Stream Sediments and Processes Controlling Element Mobility, Rough and Ready Creek, and Oregon Caves National Monument and Vicinity, Southwestern Oregon. USGS Open-File Report 98-201.

Milton, N.M., and T.L. Purdy. 1988. Response of selected plant species to nickel in western North Carolina. Castanea 53: 207–214.

Misra, K.C., and F.B. Keller. 1978. Ultramafic bodies in the southern Appalachians: A review. American Journal of Science 278: 389–418.

Mittwede, S.K. 1989. The Hammet Grove metaigneous suite: A possible ophiolite in the northwestern South Carolina Piedmont. Geological Society of America, Special Paper 231: 45–62.

Mohr, J.A., C. Whitlock, and C.N. Skinner. 2000. Postglacial vegetation and fire history, eastern Klamath Mountains, California. The Holocene 10: 587-601.

Monger, J.W.H. 1977. Ophiolitic assemblages in the Canadian cordillera. Oregon Department of Geology and Mineral Industries, Bulletin 95: 74-92.

Moore, T.R., and R.C. Zimmerman. 1977. Establishment of vegetation on serpentine asbestos mine wastes, southeastern Québec, Canada. Journal of Applied Ecology 14: 589–599.

Morford, S.L., B.Z. Houlton, and R.A. Dahlgren. 2016. Direct quantification of long-term rock nitrogen inputs to temperate forest systems. Ecology 97: 54-64.

Morrill, P.L., J.G. Kuenen, O.J. Johnson, S. Suzuki, A. Rietze, A.L. Sessions, M.L. Fogel, and K.H. Nealson. 2013. Geochemistry and geobiology of a present-day serpentinization site in California: The Cedars. Geochimica et Cosmochimica Acta 109: 222-240.

Mosher, S. 1998. Tectonic evolution of the southern Laurentian Grenville orogenic belt. Geological Society of America, Bulletin 110: 1357–1375.

Motito Marín, A., K. Mustelier, M.E. Potrony, and A. Vicario. 2004. Caracterización de la brioflora de las áreas ultramáficas cubanas. Pages 19-23. *in* R.S. Boyd, A.J.M. Baker, and J. Proctor (editors). Ultramafic Rocks: their Soils, Vegetation, and Fauna. Science Reviews, St.

Albans, Herts, UK.
Mount, H.R., and W.C. Lynn. 2004. Soil Survey Laboratory Data and Soil Descriptions for Puerto Rico and the U.S. Virgin Islands. USDA Natural Resources Conservation Service, Soil Survey Investigations Report No. 49, National Soil Survey Center, Lincoln NE.
Navarrete Gutíérrez, D.M., M-N. Pons, J.A. Calves Sánchez, and G. Echevarria. 2018. Is metal hyperaccumulation occurring in ultramafic vegetation of central and southern Mexico? Ecological Research 33: 641-649. DOI 10 1007/s11284-018-1574-4
Nelson, J.L., M. Colpron, and S. Israel. 2013. The Cordillera of British Columbia, Yukon, and Alaska: Tectonics and Metallogeny. Society of Economic Geologists, Special Publication 17: 53-103.
Nicholson, C., C.C. Sorlien, T. Atwater, J.C. Crowell, and B.P. Luyendyke. 1994. Microplate capture, rotation of the western Transverse Ranges, and initiation of the San Andreas transform as a low-angle fault system. Geology 22: 491-495 .
Nixon, E.S., and C. McMillan. 1964. The role of soil in the distribution of four grass species in Texas. American Midland Naturalist 71: 114–140.
Nixon, G.T., J.L. Hammack, G.H. Ash, L.J. Cabri, G. Case, J.N. Connelly, L.M. Heaman, J.H.G. Laflamme, C. Nuttall, W.P.E. Paterson, and R.H. Wong. 1997. Geology and Platinum-Group Element Mineralization of Alaskan-Type Ultramafic-Mafic Complexes in British Columbia. Bulletin 93, Geological Survey, Energy and Minerals, Ministry of Employment and Investment, British Columbia, Victoria.
Norris, R.M., and R.W. Webb. 1990. Geology of California. Wiley, New York.
O'Dell, R.E., and N. Rajakaruna. 2011. Intraspecific variation, adaptation and evolution. Pages 97-137 *in* S. Harrison and N. Rajakaruna. Serpentine, The Evolution and Ecology of a Model System. University of California Press, Berkeley.
O'Dell, R.E., J.J. James, and J.S. Richards. 2006. Congeneric serpentine and nonserpentine shrubs differ more in leaf Ca:Mg than in tolerance of low N, low P, or heavy metals. Plant and Soil 280: 49-64.
Ogg, C.M., and B.R. Smith. 1993. Mineral transformations in Carolina Blue Ridge-Piedmont soils weathered from ultramafic rocks. Soil Science Society of America, Journal 57: 461–472.
O'Hanley, D.S. 1996. Serpentinites: Recorders of Tectonic and Petrological History. Oxford University Press, New York.
O'Hare, J.P., and B.G. Hallock. 1980. Soils Survey of Los Padres National Forest Area, California. USDA Forest Service, Santa Barbara, CA.
Orme, D.A., and K.D. Surpless. 2019. The birth of a forearc: The basal Great Valley Group. California, USA. Geology 47: 757-761.
Ornduff, R. 2008. Thomas Jefferson Howell and the first Pacific Northwest flora. Kalmiopsis 15: 32-41
Orr, E.L., and W.N. Orr. 1996. Geology of the Pacific Northwest. McGraw Hill, New York.
Ortega-Gutierrez, F., J. Ruiz, and E. Centero-Garcia. 1995. Oaxaquia, a Proterozoic micro-continent accreted to North America during the late Paleozoic. Geology 23: 1127-1130.
Ortiz-Hernández, L.E., J.C. Escamilla-Casas, K. Flores-Castro, M. Ramírez-Cardona, and O. Acevedo-Sandoval. 2006. Caracteristicas geológicas y potential metalogenético de los principales complejos utramáficos-máficos de Mexico. Boletin de la Sociedad Geológica de

Mexico 57: 161-181.
Page, Ben. 1959. Geology of the Candelaria Mining District, Mineral County, Nevada. Nevada Bureau of Mines, Bulletin 56.
Page, B.M., and L.L. Tabor. 1967. Chaotic structure and decollement in Cenozoic rocks near Stanford University, California. Geological Society of America, Bulletin 78: 1-12.
Page, B.M., L.A. DeVito, and R.G. Coleman. 1999. Tectonic emplacement of serpentine southwest of San Jose, California. International Geology Review 41: 494-505.
Pampeyan, E.H. 1963. Geology and mineral deposits of Mount Diablo, Contra Costa County, California. California Division of Mines and Geology, Special Report 80.
Panaccione, D.G., N. Sheets, S. Miller, and J. Cummings. 2001. Diversity of *Cenococcum geophilum* isolates from serpentine and non-serpentine soils. Mycologia 93: 645–652.
Paquet, H., and G. Millot. 1972. Geochemical evolution of clay minerals in the weathering products and soils of Mediterranean climates. International Clay Conference, 4th, Proceedings: 199-206.
Pardo, G. 2009. Geology of Cuba. American Association of Petroleum Geologists, Tulsa, OK.
Parisio, S. 1981. The genesis and morphology of a serpentine soil on Staten Island. Staten Island Institute of Arts and Sciences, Proceedings 31: 2–17.
Patton, W.W., S.E. Box, and D.J. Grybeck. 1994. Ophiolites and other mafic-ultramafic complexes in Alaska. Pages 671-686 *in* The Geology of North America, volume G-1, The Geology of Alaska. Geological Society of America, Boulder, CO.
Pecora, W.T., and S.W. Hobbs. 1942. Nickel deposit near Riddle, Douglas County, Oregon. US Geological Survey, Bulletin 931-I.
Peech, M. 1965. Exchange acidity. Pages 905-913 *in* CA Black (editor). Methods of Soil Analysis, Part 2, Chemical and Microbiological Properties. American Society of Agronomy, Madison WI.
Phipps, S.P. 1984. Ophiolitic olistostromes in the basal Great Valley sequence, Napa County, northern California Coast Ranges. Geological Society of America, Special Paper 198: 103-125.
Pilbeam, D.J., and P.S. Morley. 2007. Calcium. Pages 120-144 *in* A.L. Barker and D.J. Pilbeam (editors). Handbook of Plant Nutrition, CRC Press, Boca Raton FL.
Pollard, A.J. 2016. Heavy metal tolerance and accumulation in plants of the southeastern United States. Castanea 81(4): 257-269.
Porter, S.C., and T.W. Swanson. 1998. Advance and retreat rate of the Cordilleran Ice Sheet in southeastern Puget Sound region. Quaternary Research 50: 205-213.
Pope, N., T.B. Harris, and N. Rajakaruna. 2010. Vascular plants of adjacent serpentine and granite outcrops on the Deer Isles, Maine, USA. Rhodora 112: 105-141.
Proenza, J.A., J.F. Lewis, S. Galí, E. Tauler, M. Labrador, J. Carles Melgarejo, F. Longo, and G. Bloise. 2008. Garnerite mineralization from Falcondo Ni-laterite deposit (Dominican Republic). Revista de la Sociedad Española de Mineralogia 9: 197-198.
Proenza, J.A., A. García-Casco, J.F. Lewis, S. Galí, E. Tauler, M. Labrador, and F. Longo. 2009. Textural relations and mineral compositions of garnerite ores using X-ray images. Revista de la Sociedad Española de Mineralogia 11: 155-156.
Prostka, H.J. 1962. Geology of the Sparta Quadrangle, Oregon. Oregon Department of Geology and Mineral Industries. Portland OR.

Pyke, D.R., A.J. Naldrett, and O.R. Eckstrand. 1973. Archean ultramafic flows in Munro Township, Ontario. Geological Society of America 84; 955-978.

Quick, J.E. 1981. Petrology and Petrogenesis of the Trinity Peridotite, Northern California. PhD thesis, California Institute of Technology, Pasadena CA.

Rabenhorst, M.C., and J.E. Foss. 1981. Soil and geologic mapping over mafic and ultramafic parent materials in Maryland. Soil Science Society of America, Journal 45: 1156–1160.

Rabenhorst, M.C., J.E. Foss, and D.S. Fanning. 1982. Genesis of Maryland soils formed from serpentinite. Soil Science Society of America, Journal 46: 607–616.

Radelli, L. 1989. The ophiolites of Calmalli and the Ovidada nappe of northern Baja California and west-central Sonora, Mexico. Pages 79-85 in P.L. Abbot (editor). Geological Studies in Baja California. Society of Economic Paleontologists and Mineralogists, Pacific Section, Los Angeles CA.

Radford, E. 1948. The vascular flora of the olivine deposits of North Carolina and Georgia. Journal of the Elisha Mitchell Scientific Society 64:45–106.

Raiche, R. 2009. Cedars: Sonoma County's hidden treasure. Fremontia 37(2): 3-15.

Rajakaruna, N., and B.A. Bohm. 1999. The edaphic factor and patterns of variation in *Lasthenia californica* (Asteraceae). American Journal of Botany 86: 1576-1596.

Rajakaruna, N., T.B. Harris, and E.B. Alexander. 2009. Serpentine geoecology of eastern North America: A review. Rhodora 111:21–108.

Rajakaruna, N., K. Knudsen, A.M. Fryday, R.E. O'Dell, N. Pope, F.C. Olday, and S. Woolhouse. 2012. Investigations of the importance of rock chemistry for saxicolous lichen communities of the New Idria serpentinite mass, San Benito County, California, USA. The Lichenologist 44: 695-714.

Rajakaruna, N., R.S. Boyd, and T.B. Harris (editors). 2014. Plant Ecology and Evolution in Harsh Environments. Nova Science Publishers, New York.

Ramp, L. 1978. Investigations of nickel in Oregon. Oregon Department of Geology and Mineral Industries, Miscellaneous Paper 20: 1-60.

Ravalo, E.J., M.R. Goyal, and C.R. Almodóvar. 1986. Average monthly and annual rainfall distribution in Puerto Rico. University of Puerto Rico, Journal of Agriculture 70(4): 267-275.

Rebman, J.P., and N.C. Roberts. 2012. Baja California Plant Field Guide. San Diego Natural History Museum. San Diego CA.

Reed, C.F. 1986. Flora of the Serpentinite Formations in Eastern North America, with Descriptions of the Geomorphology and Mineralogy at the Formations. Contributions of the Reed Herbarium 30. Reed Herbarium, Baltimore, MD.

Reed, W.P. 2006. Soil Survey of Colusa County, California. USDA, Natural Resources Conservation Service, Washington, DC.

Reeves, R.D., R.M. Macfarlane, and R.R. Brooks. 1983. Accumulation of nickel and zinc in western North American genera containing serpentine-tolerant species. American Journal of Botany 70: 1297-1303.

Reeves, R.D., A.J.M. Baker, A. Borhidi, and R. Berazain. 1998. Nickel hyperaccumulation in the flora of Cuba. Annals of Botany 83: 29-38.

Reeves, R., A.J.M. Baker, R Romero. 2007. The ultramafic flora of the Santa Elena peninsula, Costa Rica: A biochemical reconnaissance. Journal of Geochemica Exploration 93: 153-159.

Reid-Soukup, D.A., and A.L. Ullery. 2002. Smectites. Pages 467-499 *in* J.B. Dixon and D.G. Schulze (editors), Soil Mineralogy with Environmental Applications. Soil Science Society of America, Madison, WI.

Rice, J.S. 1957. Nickel. California Division of Mines and Geology, Bulletin 176: 391-399.

Rice, J.S., and G.B. Cleveland. 1955. Laterite silicification of serpentinite in the Sierra Nevada, California. Geological Society of America, Bulletin 66: 1660.

Rice, J.S., and D. Wagner. 1991. Geology and mineralogy of Ring Mountain: A popular nature reserve. California Geology 44: 99-106.

Roberts, B.A. 1980. Some chemical and physical properties of serpentine soils from western Newfoundland. Canadian Journal of Soil Science 60: 231–240.

Roberts, B.A. 1992. Ecology of serpentinized areas, Newfoundland, Canada. Pages 75–113 *in* B.A. Roberts and J. Proctor (editors). The Ecology of Areas with Serpentinized Rocks: A World View. Kluwer, Dordrecht.

Roberts, R.C. 1936, no. 8, issued in 1942. Soil Survey of Puerto Rico. USDA, Bureau of Plant Industry, Division of Soil Survey.

Ross, G.J., and H. Kodama. 1976. Experimental alteration of a chlorite into a regularly inter-stratified chlorite-vermiculite by chemical oxidation. Clays and Clay Minerals 22: 205-211.

Ruckmick, J.C., and J.A. Noble. 1959. Origin of the ultramafic complex at Union Bay southeastern Alaska. Geologic Society of America, Bulletin 70: 981-1018.

Ruiz J., P.F. Patchett, and F. Ortega-Gutierrez. 1988. Proterozoic and Phanerozoic basement terranes of Mexico from Nd isotope studies. Geological Society of America, Bulletin 100: 274-281.

Ryan, B.D. 1988. Marine and maritime lichens on serpentine rocks on Fidalgo Island, Washington. The Bryologist 91: 186-190.

Safford, H. 2011. Serpentine endemism in the California flora. Fremontia 38(4)-39(1): 32-39.

Safford, H,, and S. Harrison. 2001. Ungrazed road verges in a grazed landscape: interactive effects of grazing, invasion, and substrate on grassland diversity. Ecological Applications 11: 1112-1122.

Safford, H., and S. Harrison. 2004. Fire effects on plant diversity in serpentine versus sandstone chaparral. Ecology 85: 539-548.

Safford, H,, and E,D, Miller. 2020. An updated database of serpentine endemism in the California flora. Madrono 67; 85-104.

Safford, H.D., J.S. Viers, and S.P. Harrison. 2005. Serpentine endemism in the California flora: A database of serpentine affinity. Madroño 52(4): 222-257.

Saint John, H. 1963. Flora of Southeastern Washington. Outdoor Pictures, Escondido CA.

Saleeby, J. 2011. Geochemical mapping of the Kings-Kaweah ophiolite body, California–evidence for progressive mélange formation in a large offset transform-subduction initiation environment. Geological Society of America, Special Paper 490: 31-74.

Sanborn, P. 2013. Central interior soil lithology study: soil descriptions and characterization data. University of Northern British Columbia, Prince George.

Schulze, D.G. 1989. An Introduction to Soil Mineralogy. Pages 1-34 *in* Minerals in Soil Environments. Soil Science Society of America, Madison, WI.

Schwartz, J.J., A. Snoke, C. Frost, C. Barnes, L. Gromet, K. Johnson. 2010. Analysis of the Wallowa-Baker terrane boundary: Implications for tectonic accretion in the Blue Mountains province, northeastern Oregon. Geological Society of America, Bulletin 122: 517-536.

Seabloom, E.W., E. Borer, V. Boucher, R. Burton, K. Cottingham, L Goldwasser, W. Gram B. Kendall, and F. Micheli. 2003. Competition, seed limitation, disturbance, and reestablishment of California native annual forbs. Ecological Applications 13: 575-592.

Sedlock, R.L. 2003. Geology and tectonics of the Baja California peninsula and adjacent areas. Geological Society of America, Special Paper 374: 1-42.

Sedlock, R.L., F. Ortega-Gutiérrez, and R.C. Speed. 1993. Tectonostratigraphic Terranes and Tectonic Evolution of Mexico, Geological Society of America, Boulder CO.

Ségalen, P., D. Bosch, A. Cardenas, E. Camacho, A. Bouleau, H. Guénin, and D. Hamboud. 1980. Aspects minéralogiques et pédogénétiques de deux sols dérivés de péridotites dans l'oeste de Cuba. Cahiers ORSTOM, série Pédologie 18: 273-284.

Ségalen, P., M. Gautheyrou, H. Guénin, E. Camacho, D. Bosch, and A Cardenas. 1983. Étude d'un sols dérivés de péridotite dans l'oeste de Cuba, aspects physiques et chimiques. Cahiers ORSTOM, série Pédologie 20: 239-245.

Senkayi, A.L. 1977. Clay mineralogy of poorly drained soils derived from serpentinite rocks. University of California, Davis. PhD Dissertation.

Shacklette, H.T., and J.G. Boerngen. 1984. Element Concentrations in Soils and Other Surficial Materials of the Conterminous United States. US Geological Survey, Professional Paper 1270.

Shakesby, R.A, 1997. Pronival (protalus) ramparts: A review of forms, processes, diagnostic criteria, and paleoenvironmental implications. Progress in Physical Geography 21: 394-318.

Shannon, R.D. 1976. Revised effective ionic radii and systematic studies of interatomic distances in halides and chalcogenides. Acta Crystallographica A32: 751-767.

Sharp, R.P. 1960. Pleistocene glaciation in the Trinity Alps of northern California. American Journal of Science 258: 305-340.

Shreve, F. 1951. Vegetation of the Sonoran Desert. Carnegie Institute of Washington, Publication 591, 192 pages.

Sikes, K., D. Roach, and J. Buck. 2013. Classification and Mapping of Vegetation from Three Fen Sites of the Shasta-Trinity National Forest, California. California Native Plant Society, Sacramento, CA.

Silberling, N.J., D.L. Jones, J.W.H. Monger, and P.J. Coney. 1992. Lithotectonic terrane map of the North American Cordillera. USGS Miscellaneous Investigation Map I-2176.

Simmons, C.S., J.M. Tárano T., and J.S. Pinto Z. 1959. Clasificación de Reconocimiento de los Suelos de la República de Guatemala. Instituto Agropecuario Nacional, Ministerio de Agricultura. Guatemala.

Singer, A. 2002. Palygorskite and sepiolite. Page 555-583 *in* J.B. Dixon and D.G. Schulze. Soil Mineralogy with Environmental Applications. Soil Science Society of America, Madison, WI.

Sirois, L., and M. Grandtner. 1992. A phyto-ecological investigation of the Mount Albert serpentine plateau. Pages 115–133 *in* B.A. Roberts and J. Proctor (editors). The Ecology of Areas with Serpentinized Rocks: A World View. Kluwer, Dordrecht, The Netherlands.

Sirois, L., F. Lutzoni, and M. Grandtner. 1988. Les lichens sur serpentine et amphibolite du plateau du Mont Albert, Gaspésie, Québec. Canadian Journal of Botany 66: 851–862.

Sivarajasingham, S. 1961. Weathering and soil formation on ultrabasic and basic rocks under humid tropical conditions. PhD dissertation, Cornell University, Ithaca, NY.

Smith, C.A.S., J.C. Meikle, and C.F. Roots (editors). 2004. Ecoregions of the Yukon Territory: Bio-physical properties of Yukon landscapes. Agriculture and Agri-food Canada. PARC Technical Bulletin No. 04-01. Summerland, BC.

Smith, C.F. 1998. A flora of the Santa Barbara region, California: an annotated catalog of the native, naturalized, and adventive vascular plants of mainland Santa Barbara County, adjacent related areas, and four nearby Channel Islands. Santa Barbara Botanic Garden & Capra Press.

Smith, D.W., and W.D. Broderson. 1989. Soil Survey of Lake County, California. US Government Printing Office, Washington, DC.

Snow, C.A. 2002. Cuesta Ridge ophiolite: New field & geochemical evidence for origin & evolution of the Coast Range ophiolite, California. Geological Society of America, Cordillera Section, 98th Annual Meeting, Corvallis, OR, Session 11, abstract.

Soil Classification Working Group. 1998. The Canadian System of Soil Classification. Agriculture and Agri-Food Canada, Ottawa, ON, Canada. Publication 1646.

Soil Survey Staff. 1993. Soil Survey Manual. Agriculture Handbook No. 18. US Government Printing Office, Washington, DC.

Soil Survey Staff. 1999. Soil Taxonomy: A Basic System for Making and Interpreting Soil Surveys. USDA, Agriculture Handbook No. 436. US Government Printing Office, Washington, DC.

Soil Survey Staff. 2014. Keys to Soil Taxonomy. USDA, Natural Resources Conservation Service.

Southworth, D., L.E. Tackaberry, and H.B. Massacotte. 2013. Mycorrhizal ecology on serpentine soils. Plant Ecology and Diversity 7: 445-455.

Stebbins, G.L. 1984. Serpentine flora: notes on prominent sites in California, the northern Sierra Nevada. Fremontia 11(5): 26-28.

Terlizza, D.E., and E.P. Karlander. 1979. Soil algae from a Maryland serpentine formation. Soil Biology and Biochemistry 11:205–207.

Thomas, D.J. 1998. Soil survey of Clay County, North Carolina. USDA, Natural Resources Conservation Service, Washington, DC.

Thompson, G.A., and B. Robinson. 1975. Gravity and magnetic investigation of the Twin Sisters dunite, northern Washington. Geological Society of America, Bulletin 86: 1413-1422.

Thorne, R.F. 1967. A Flora of Santa Catalina Island. Aliso 6(3): 1-77.

Tollo, R.P., L. Corriveau, J.M. McLelland, and M.J. Bartholemew. 2004. Proterozoic tectonic evolution of the Grenville Orogen in North America. Geological Society of America Memoir 197: 1-19.

Trescases, J-J. 1975. L'évolution géochemique supergène des roches ultrabasiques en zone tropicale - formation des gesements nickélèferous de Nouvelle Calédonie. Mémoires ORSTOM, No. 78, 259 pages, photographic plates, English abstract.

Trescases, J-J. 1997. The lateritic nickel-ore deposits. Pages 125-138 *in* H. Paquet and N. Clauer (editors). Soils and Sediments. Springer, Berlin.

Troll, C. 1971. Landscape ecology (geoecology) and biogeocoenology–a terminological study. Geoforum 8: 43-46.

Turekian, K.K., and K.H. Wedepohl. 1961. Distribution of the elements in some major units of the earth's crust. Geological Society of America Bulletin 72: 175-192.

Tyndall, R.W. 1992. Historical considerations of conifer expansion in Maryland serpentine “barrens.” Castanea 57: 123-131.
Tyndall, R.W. 1994. Contributions of the barrens symposium, preface. Castanea 59: 182–183.
Unruh, J.R., T.A. Dumitru, and T.L. Sawyer. 2007. Coupling of early Tertiary extension in the Great Valley forearc basin with blueschist exhumation in the underlying Franciscan accretionary wedge of Mount Diablo. Geological Society of America, Bulletin 119: 1347-1367.
Vallier, T.I., and H.C. Brooks (editors). 1995. Geology of the Blue Mountains Region of Oregon, Idaho, and Washington: Petrology and Tectonic Evolution. US Geological Service, Professional Paper 1438.
Valls, R.A. 2006. Geology and Geochemical Evolution of the Ophiolitic Belts in Guatemala. A Field Guide to Nickel Bearing Laterites, 6th edition. Nichromet Extraction Inc., Toronto.
van der Ent, A., A.J.M. Baker, R. Reeves, A. Pollard, and H. Schat. 2013. Hyperaccumulators of metals and metaloid trace elements: facts and fiction. Plant and Soil 362: 319-334.
Vedder, J.G., D.G. Howell, and H. McLean. 1980. Stratigraphic and structural relations to pre-Tertiary rocks on the perimeter of the Santa Maria basin. AAPG Bulletin 64: 450 (abstract).
Velde, B., and A. Meunier. 2008. The Origin of Clay Minerals in Soils and Weathered Rocks. Springer-Verlag, Berlin.
Villanova-de-Benavent, C., F. Nieto, J.A. Proenza, and S. Gali. 2011. Talc-serpentine-like “garnierites” from Falcondo Ni-laterite deposit (Dominican Republic): a HRTEM approach. Revista de la Sociedad Española de Mineralogia 15: 197-198.
Wahrhaftig, C. 1966. Physiographic divisions of Alaska. US Geological Survey, Professional Paper 482.
Wahrhaftig, C. 1984. A Streetcar to Subduction. Am. Geophysical Union, Washington, DC.
Wahrhaftig, C., and J.S. Birman. 1965. The Quaternary of the Pacific Mountain system in California. Pages 299-340 in H.E. Wright, Jr., and D.G. Frey (editors). Princeton University Press, Princeton, N.J.
Wakabayashi, J. 2013. Paleochannels, stream incision, erosion, topographic evolution, and alternative explanations of paleoaltimetry, Sierra Nevada, California. Geosphere 9:191-215.
Wakabayashi, J. 2015. Anatomy of a subduction complex: architecture of the Franciscan complex, California, at multiple length and time scales. International Geology Review 57: 1-78.
Wakabayashi, J., and Y. Dilek. 2003. What constitutes “emplacement”of ophiolite? Mechanisms and relationships to subduction initiation and formation of metamorphic soles. Pages 427-447 in Y. Dilek and P.T. Robinson (editors). Ophiolites in Earth History. Geological Society, London, Special Paper 218.
Wakabayashi, J., and T.L. Sawyer. 2001 Stream incision, tectonics, uplift, and evolution of topography of the Sierra Nevada, California. Journal of Geology 109: 539-562.
Walker, G.W., and N.S. MacLeod. 1991. Geologic Map of Oregon. US Geological Survey.
Walter, H. Vegetation of the Earth in Relation to Climate and Eco-physiological Conditions. translated by J. Wieser. Springer-Verlag.
Weiss, S.B. 1999. Cars, cows, and checkerspot butterflies: nitrogen deposition and management of nutrient-poor grasslands for a threatened species. Conservation Biology 13: 1476-

1486.
Wells, P.V. 1962. Vegetation in relation to geological substratum and fire in the San Luis Obispo quadrangle, California. Ecological Monograph 32: 79-103.
Wheeler, A.G. 1988. *Diabrotica cristata*, a chrysomelid (Coleoptera) of relict Midwestern prairies discovered in eastern serpentine barrens. Entomological News 99: 134–142.
Whipple, J., and E. Cole. 1979. An ecological survey of the proposed Mount Eddy Research Natural Area. USDA Forest Service, Pacific Southwest Research Station, Berkeley, CA.
White, G.N., and J.B. Dixon. 2002. Kaolin-serpentine minerals. Pages 389- 414 *in* J.B. Dixon and D.G. Schultz. Soil Mineralogy with Environmental Applications. Soil Science Society of America, Madison, WI.
Whittaker, R.H. 1960. Vegetation of the Siskiyou Mountains, Oregon and California. Ecological Monographs 30: 279-338.
Wiggins, I.L. 1980. Flora of Baja California. Stanford University Press, Stanford, CA.
Wildman, W.E., M.L. Jackson, and L.D. Whittig. 1968. Iron-rich montmorillonite formation in soils derived from serpentinite. Soil Science Society of America, Proceedings 32: 787-794.
Williams, H., and P. St-Julien. 1982. The Baie Verte-Brompton Line: Continent-ocean interface in the Northern Appalachians. Geological Association on Canada, Special Paper 24: 177–207.
Williamson, J.N., and S. Harrison. 2002. Biotic and abiotic limits to the spread of exotic re-vegetation species in oak woodland and serpentine habitats. Ecological Applications 12: 40-51.
Wilson, J.L., D.R. Ayres, S. Steinmaus, and M. Baad. 2010. Vegetation and flora of a biodiversity hotspot: Pine Hill, El Dorado County, California, USA. Madroño 56: 246-278.
Wilson, M.J., and M.L. Berrow. 1978. The mineralogy and heavy metal content of some serpentinite soils in north-east Scotland. Chemie der Erde 37: 181-205.
Wing, L.A. 1951. Asbestos and serpentine rocks of Maine. Maine Geological Survey, Augusta, ME. Report of the State Geologist 1949–1950: 35–46.
Wlodarski, R.J. 1979. Catalina Island soapstone manufacture. California and Great Basin Anthropology, Journal 1: 331-355.
Wolf, A.T., S. Harrison, and J.L. Hamrick, 2000. The influence of habitat patchiness on genetic diversity and spatial structure of a serpentine endemic plant. Conservation Biology 14: 454-463.
Worthington, J.E. 1964. An exploration program for nickel in the southeastern United States. Economic Geology 59: 97–109.
WRB. 2014. World Reference Base for Soil Resources. Food and Agriculture Organization of the United Nations.
Wright, R.L.L., J. Nagel, and K.C. Taggart. 1982. Alpine ultramafic rocks of southwestern British Columbia. Canadian Journal of Earth Sciences 19: 1156-1173.
Yuan, T.L. 1959. Determination of exchangeable hydrogen in soils by a titration method. Soil Science 88: 164-167.
Zika, P.F., and K.T. Dann. 1985. Rare plants on ultramafic soils in Vermont. Rhodora 87: 293–304.

Index

About the Author

Earl Alexander was born and educated in Ohio, with a PhD degree from the Ohio State University. He has mapped soils, developed interpretations for soils and land management, and trained and supervised experts in these activities for the Forest Service of the US Department of Agriculture, the University of Nevada, and the Food and Agriculture Organization (FAO) of the United Nations. His professional experiences have been mostly in western North America, from Alaska to Baja California, and in Central America. He has done geoecological mapping in California, Oregon, and Central America. He is the author of a soils book, *Soils in Natural Landscapes*, and the first author of another book, *Serpentine Geoecology of Western North America.* Many of his professional articles are cited in the book.

Ultramafic Geoecology of North America
Arctic to Caribbean

iUniverse books may be ordered through booksellers or by contacting:

iUniverse
1663 Liberty Drive
Bloomington, IN 47403
www.iuniverse.com
844-349-9409

ISBN: 978-1-6632-3439-1 (sc)
978-1-6632-3440-7 (e)

Library of Congress Control Number: 2022900374

Print information available on the last page.

iUniverse rev. date: 02/01/2022

Printed in the United States
by Baker & Taylor Publisher Services